BIM Technology:
The Second Revolution in Architectural Design

BIM技术
第二次建筑设计革命

主编　欧阳东
Editor-in-Chief　**Ouyang Dong**

中国建筑工业出版社
China Architecture & Building Press

图书在版编目（CIP）数据

BIM技术——第二次建筑设计革命/欧阳东主编.
北京：中国建筑工业出版社，2013.7
ISBN 978-7-112-15557-6

Ⅰ.① B… Ⅱ.①欧… Ⅲ.①建筑设计－计算机辅助设计－应用软件 Ⅳ.①TU201.4

中国版本图书馆CIP数据核字（2013）第137659号

本书是作者近几年在BIM技术应用推广方面的实践和感悟，作者将逐渐形成的具有本土特点的BIM技术解决方案融入书中，其中有方法、案例，也有实践、经验；有问题、思考，也有破解、建议。全书共7章，分别对BIM技术的现状、趋势、优势、标准、软件、问题等的分析和总结，对我国建筑设计行业各企业实施BIM技术是很好的指导与借鉴。全书图文并茂，突出要点，实操性强，中英文对照，有较强的指导性、参考性、实用性和权威性。

本书既可供政府部门、建设单位、设计单位、施工单位、监理公司、物业公司、软件厂商、系统集成商等单位的负责人、设计师、技术人员等学习参考，也可作为大中专学校相关专业的教学参考书。

责任编辑：刘　江　范业庶
装帧设计：北京美光设计制版有限公司
责任校对：肖　剑　刘　钰

BIM技术
——第二次建筑设计革命

主编　欧阳东

*

中国建筑工业出版社出版、发行（北京西郊百万庄）
各地新华书店、建筑书店经销
北京美光设计制版有限公司制版
北京画中画印刷有限公司印刷

*

开本：880×1230毫米　1/32　印张：7 1/8　字数：200千字
2013年8月第一版　2014年7月第三次印刷
定价：80.00元　（含光盘）
ISBN 978-7-112-15557-6
(24168)

版权所有　翻印必究
如有印装质量问题，可寄本社退换
（邮政编码　100037）

编委会 Editorial Board

主编　欧阳东
教授级高级工程师/院长助理
中国建筑设计研究院（集团）

Editor-in-Chief　Ouyang Dong
Professor-level Senior Engineer / Assistant President, China Architecture Design & Research Group (CAG)

特邀编委 Guest Editorial Members

魏篙川
教授级高级建筑师/国际所所长
中国建筑设计研究院

于　洁
高级建筑师/BIM设计研究中心主任
中国建筑设计研究院

杨　旭
一级注册建筑师
中国建筑设计研究院

樊　珣
BIM讲师/建筑师
中国建筑设计研究院

秦　军
技术总监
北京互联立方技术服务有限公司

罗海涛
技术总监
美国Autodesk(中国)有限公司

俞兴扬
技术销售总监
美国Bentley 软件(北京)有限公司

游　洋
技术销售总监
芬兰Progman（中国）有限公司

黄　琨
副院长
北京理正软件股份有限公司

赵景峰
BIM事业部总经理
北京鸿业同行科技有限公司

翻　译　喻蓉霞　及炜煜　欧阳凌翔

Wei Gaochuan
Professor-level Senior Architect and Director of the International Department, China Architecture Design & Research Group (CAG)

Yu Jie
Senior Architect/Director of BIM Design & Research Center, China Architecture Design & Research Group (CAG)

Yang Xu
First Class Registered Architect, China Architecture Design & Research Group (CAG)

Fan Xun
BIM Lecturer/Architect, China Architecture Design & Research Group (CAG)

Qin Jun
Technical Director, Beijing is BIM Technical Services Co., Ltd

Luo Haitao
Technical Director, Autodesk (China) Inc.

Yu Xingyang
Technical Sales Director, Bentley (China) Intelligent Technology Co., Ltd

You Yang
Technical Sales Director, Progman (China) Co., Ltd

Huang Kun
Vice President, Beijing Leading Software Co., Ltd

Zhao Jingfeng
BIM Department Manager, Beijing Hongye Technology Co., Ltd

Translator　Yu Rongxia, Ji Weiyu, Ouyang LingXiang

作者简介
About the Author

欧阳东
中国建筑设计研究院（集团）
院长助理、教授级高工
国务院政府特殊津贴专家
集团信息化技术委员会副主任
全国智能建筑电气技术情报网常务副理事长
中国建筑节能协会建筑电气节能专业委员会常务副主任
北京市和住房和城乡建设部人才中心正高评委

Ouyang Dong
China Architecture Design & Research Group (CAG)
Assistant President, Professor-level Senior Engineer
State Council Expert with Special Allowance
Deputy Director of CAG Information Technology Committee
Executive Vice President of China Intelligent Building Technology Information Association
Executive Deputy Director of Building Electricity and Intelligent Efficiency Committee of China Association of Building Efficiency
Senior Judge of the Talent Center of the Beijing Municipality and the Ministry of Housing and Urban-Rural Development of the People's Republic of China

　　欧阳东，1982年毕业于重庆建筑工程学院自动化专业，工学学士，2009年毕业于厦门大学，EMBA高级经济管理硕士；2005年取得国家一级电气注册工程师职称。曾任综合所所长、机电院院长、运营中心主任等职务，现任集团院长助理、集团总法律顾问；社会兼职：《智能建筑电气技术》杂志副社长；"中国智能建筑信息网"理事长。

　　Mr. Ouyang Dong obtained his Bachelor's degree in Automation from Chongqing Institute of Architecture and Civil Engineering in 1982 and an Executive MBA (EMBA) in Senior Business Administration and Management from Xiamen University in 2009, and became a National First Class Registered Electrical Engineer in 2005. He was previously Director of the Comprehensive Department, President of the Electrical and Mechanical Institute, Director of the Operation Center of China Architecture Design & Research Group (CAG), and is currently Assistant to the CAG President and the Group General Legal Counsel. His appointments include Deputy Director of the publication *Intelligent Building Electrical Technology* and President of the China Intelligent Building Information website.

作为工种负责人，参与了几十项大中型项目的设计工作，并取得了"北京梅地亚中心"等多个设计项目的国家级、省部级优秀设计奖。作为项目负责人，参与了多项企业级、部级和国家级科研项目，并取得了《建筑机电设备开放式通信协议研究》等多个科研项目的住房和城乡建设部华夏建设科学技术奖二、三等奖，主编了《医疗建筑电气设计规范》（中英文）；作为第一专利人，发明了"智能型灯光面板"获得三项专利。作为主编或副主编完成了《建筑机电节能设计手册》等十几本著作的编著工作，并均已正式出版；独著《建筑机电节能设计探讨》、《设计企业管理研究》等十几篇技术论文和管理论文。主持过多次全国性行业会议，多次在行业会议上宣讲了"建筑机电节能设计研究"、"管理创新——企业发展之精髓"、"BIM技术——第二次建筑设计革命"。

作为院（集团）院长助理兼设计运营中心主任，组织完善、调整、建立了一套"新的设计组织架构体系——项目经理和设计研究室主任的强矩阵管理架构"，取得非常好的经营业绩，各项经营指标连续四年均创历史新高。作为负责人组织BIM技术应用和推广工作，并取得了BIM最佳企业应用奖和五个BIM项目最佳设计奖。曾获得院（集团）管理创新特殊贡献奖、"十一五"科技创新奖、科研管理奖。

As principal of his profession, Mr. Ouyang has participated in designing dozens of large and medium-sized projects, and several of his design projects, such as Media Center Hotel Beijing have earned national and provincial excellent design awards. As project principal, he has been involved in several corporate, ministerial and national scientific research projects. Mr. Ouyang's multiple scientific research projects like *Open Communication Protocol Study of Building Mechanical and Electrical Equipment* were awarded second and third prizes in the China Construction Science and Technology Awards from the Ministry of Housing and Urban-Rural Development of the People's Republic of China and he also edited *Code for the Electrical Design of Medical Buildings* (Chinese and English versions). In addition, Mr. Ouyang holds three patents, including one for intelligent lighting panels. As chief or associate editor, he has completed the compilation of a dozen works that have all been published, like *Building Mechanical and Electrical Efficiency Design Manual*, and a dozen of academic papers in technology and management, such as *Discussion on Building Mechanical and Electrical Efficiency Design* and *Study on Design Enterprise Management*. Moreover, he has chaired national industry conferences numerous times and presented papers on *Study on Building Mechanical and Electrical Efficiency Design, Management Innovation: the Essence of Corporate Development*, and *BIM Technology: The Second Revolution in Architectural Design*.

As Assistant President of CAG and Director of Design Operation Center, Mr. Ouyang has conducted, adjusted and established a set called *New Design Organizational Structure System: Strong Matrix Management Structure for Project Managers and Design & Research Office Directors*, and yielded outstanding business results with all business indicators hitting the record highs for four consecutive years. As principal, he has implemented the application and promotion of BIM technology, and was awarded the Best Enterprise with BIM Application and five Awards of the Best Design for BIM Projects. He was also awarded the CAG Special Services Award to Management Innovation, and 11th Five-Year Technology Innovation Award and Scientific Research Management Award.

序
Foreword

　　信息技术日新月异的发展，有力地推进了各个行业的技术水平、管理能力的提升。在建筑设计行业，建筑设计不仅是一种艺术创作，更是一个涉及多个专业领域、综合性很强的系统工程，它包含的大量信息需要强有力的技术手段去采集、分类、分析、检索和传输；而建筑信息模型BIM（Building Information Modeling）技术作为数字建筑技术中出现的新概念、新理念和新技术，将为建筑设计的进步提供强有力的技术支撑。

　　2010年1月27日，在清华大学举办的"BIM对中国建筑业未来影响及中国BIM标准的研究制定"专家讨论会上，我们提出的第一个讨论话题是"BIM对未来中国建筑行业将引发的是技术革命还是产业革命"。记得与会专家对BIM是否可以成为产业革命还将信将疑，今天我们高兴的看到，该书作者已经从产业实际发展的角度，提出了"BIM技术是第二次建筑设计革命"，这将引发建筑设计行业从技术手段到商业模式等各层面上的颠覆性变革。

　　作为我国建筑设计行业的领跑者，中国建筑设计研究院也是BIM技术的先行先试先用者。作者结合近几年在BIM技术应用推广方面的实践和感悟，将逐渐形成的具有本土特点的BIM技术解决方案融入书中，其中有方法、案例，也有实践、经验；有问题、思考，

　　The development of information technologies vigorously promotes the technology level and ascends management ability in a wide range of industries. In architectural design industry, design is not only a kind of artistic creation, but also a very strong comprehensive system engineering related to multiple professional areas, which contains lots of information needing strong technical method to collect, classify, analysis, retrieve and transmit. BIM (Building Information Modeling) technology, as a new concept, new idea and new technology in digital building technology, provides strong technical support to the progress of architecture design.

　　On January 27, 2010, in an expert discussion conference held in Tsinghua University which named "The impact of BIM on future Chinese construction industry and BIM standards formulation", we put forward the first topic " Whether a technology revolution or industrial revolution for future Chinese architectural industry that will be triggered by BIM ? ". I remember experts were skeptical about BIM at that time, however, today we are happy to see that, the author has put forward "BIM technology is the second architecture design revolution " from the perspective of actual development, which will trigger an disruptive change from technology to the business model and so on various levels in architectural design industry.

　　As a leader of Chinese architectural design industry, CAG is also a forerunner of BIM technology. Considering the BIM application experience and comprehension in recent years, the author composites a localized characteristic of BIM technology solutions into the book, including methods, case studies, also the practice and experience; besides, there are problems, thinking, cracking, and advice. Especially the analysis and summary of BIM's

也有破解、建议。特别是书中对BIM技术的现状、趋势、优势、标准、软件、问题等的分析和总结,对我国建筑设计行业各企业实施BIM技术是很好的指导与借鉴。

BIM技术的推广和普及,不仅需要政府的积极引导,也需要更多像本书作者这样的实践者与热心人,愿意将其实践经验与社会各界分享。这体现了一个科技工作者的社会责任。在此,我向他们表示致敬!

中国工业化与信息化融合的道路刚刚起步,任重而道远。希望本书的出版对指导建筑设计企业实施基于BIM技术的信息化战略起到积极作用,促进BIM技术在建筑设计行业的广泛理解与深入应用,推动中国工程建设行业在我国绿色城镇化的快速、健康发展中朝阳永铸。

current situation, trend, advantage, standards, software, or problem, for the architectural design industry are good guides and references for the implementation of BIM technology in our Chinese enterprise

The promotion and popularization of BIM technology, not only needs the government's active guide, but also need more practitioners and enthusiasts, like author willingly to share practical experience to the society, which presents social responsibility of a scientific and technical workers. Here, I pay my respect to them!

The path to the fusion of industrialization and information has just started, and a long way to go. I Hope that the publication of the book could play a positive role to guide the architectural design enterprises to carry out the information strategy based on BIM technology, promoting the broad understanding and in-depth application of BIM technology in architectural design industry, motivating rapid and healthy development of Chinese construction industry in China's green urbanization.

中国工程院院士
2013-7-1
Sun Jiaguang
Academician of the Chinese Academy of Engineering
JUL 1, 2013

前言
Preface

形势 Trend	党的十八大胜利召开，回顾"十一五"（2006–2011）时期取得的辉煌业绩，国民生产总值（GDP）从26.6万亿元增加到51.9万亿元，跃升到世界第二位，年增速9.3%；科技投入5年累计8729亿元，年增速超过18%；提出了科技驱动创新和生态节能环境，并明确提出了实现"工业化、信息化、城镇化"。因此，科技创新和节能环保将是中国建筑业永恒主题。 With its successful assembly, the CPC's 18th National Congress recalled the brilliant achievements obtained during the "11th Five-Year Plan" (2006–2011), in which China's GDP had an annual growth rate of 9.3%, an increase from 26.6 trillion Yuan to 51.9 trillion Yuan and the economy ranked second in the world, and investment in science and technology amounted to 872.9 billion Yuan in five years, with an annual growth rate of more than 18%. Meanwhile, science and technology-driven by innovation and eco-energy environment were proposed, and it was made clear to achieve industrialization, informationization and urbanization. Therefore, technological innovation, energy saving and environmental protection will be ongoing themes pursued by China's construction industry.
目的 Goal	贯彻执行国家技术经济和节能政策，在中国建筑行业推广BIM新技术，促进行业科技进步。BIM技术的全面应用将显著提高工程项目设计、施工、运营全生命周期的质量及效率，降低成本，提升集成化程度，并产生巨大的经济和社会效益，也是实现项目精细化管理、企业集约化经营的有效途径。 To implement the national technological economy and energy-saving policy, promote new technologies including BIM technology in China's construction industry, and advance industrial scientific and technological progress. The full application of BIM technology will significantly improve the quality and efficiency of project design, construction and full operation life cycle, reduce costs, enhance the degree of integration and generate immense economic and social benefits, and also, serve as an effective means to realize the project's detailed management and intensive corporate management.

(续表)

主题 Theme	在中国建筑设计行业第一个提出"BIM技术——第二次建筑设计革命",必将引发建筑设计行业从技术手段到商业模式等所有层面上颠覆性变化;同时明确指出"通过改变建筑设计模式,可以提高企业竞争力",BIM已经不是设计企业发展的可选项,而是必选项,设计企业投资BIM是投资企业未来的最佳途径之一。这场建筑设计革命将不受个人好恶和思维习惯的束缚而向前推进,谁愈先采用,谁愈早受益。 The first introduction of "BIM Technology: The Second Revolution in Architectural Design" in China's architecture industry will inevitably trigger a fundamental change at all levels, from technological means to business modes throughout the industry, but this will also enhance corporate competiveness through changing architecture design modes, and so BIM is no longer an option in the development of design enterprises, but a must. Therefore, investing in BIM is one of the optimal paths for investing in an enterprise's future. This architectural design revolution will, regardless of personal tastes and thinking habits, will move everything forward. Whoever adopts it earlier will benefit sooner.
需求 Needs	1. 经济发展的需求（GDP、生产力、可持续健康发展等）； 2. 技术进步的需求（手段、流程、质量、效率等）； 3. 核心竞争力需求（人才、品牌、战略等）； 4. 行业发展的需求（科技研发、节约投资等）； 5. 城市发展的需求（城镇化、数字城市等）； 6. 社会进步的需求（节能环保、社会责任等）。 1. Economic development needs (GDP, productivity, sustainable and healthy development, etc.); 2. Technological progress needs (means, processes, quality and efficiency, etc.); 3. Core competitiveness needs (human resources, brand and strategy, etc.); 4. Industrial development needs (research and development and investment savings, etc.); 5. Urban development needs (urbanization and digital city, etc.); 6. Social progress needs (energy saving and environmental protection, social responsibility, etc.).

（续表）

变革 Revolution	1. 传统思维的变革（习惯意识，传统想法等）； 2. 技术手段的变革（科技创新、科研转化生产力、软件技术变革等）； 3. 商业模式的变革（开拓新的经营模式及市场、传统工作模式等）； 4. 城镇化建设变革（政府审批、管理运营的统筹、应急预案等）； 5. 产业信息的变革（全生命周期、信息的全过程传递和综合使用等）； 6. 建筑产业的变革（信息化水平、工业化水平等）。 1. Revolutionizing traditional thinking (habitual awareness and traditional ideas, etc.); 2. Revolutionizing technical means (scientific and technological innovation, scientific research converted to productivity and changes in software technology, etc.); 3. Revolutionizing business modes (to develop new business modes, markets and traditional operation modes, etc.); 4. Revolutionizing urbanization (government approval, planning of operation management and contingency plans, etc.); 5. Revolutionizing industrial information (full life cycle, entire process of delivery and integrated use of information, etc.); 6. Revolutionizing the construction industry (informationization level and industrialization level, etc.).
内容 Content	1. BIM技术的现状和未来发展趋势（含中国三届BIM设计大赛）； 2. BIM技术应用的优势和劣势比较； 3. BIM技术标准的现状和编制规划； 4. BIM技术在企业应用方面的推广（含BIM设计培训和协同设计）； 5. BIM技术在设计项目方面的推广（含BIM设计项目流程）； 6. BIM技术设计应用软件汇总； 7. BIM技术应用遇到的问题和政府影响。 1. Current Status and Future Trend of BIM Technology (including the three contests of "Innovation Cup" for BIM Design); 2. Comparison of Application of the Advantages and Disadvantages of BIM Technologg; 3. Current Status and Compiling Planning of BIM Technology Standards; 4. BIM Promotion in Enterprise Application (including BIM design training and collaborative design); 5. BIM Promotion in Design Projects (including BIM design project process); 6. Summary of BIM Technology in Design Application Software; 7. Problems in BIM Technology Implementing and the Role of Government.

（续表）

对象 Object	政府部门、建设单位、设计单位、施工单位、监理公司、物业公司、软件厂商、系统集成商等单位的负责人、设计师、技术人员等。 Directors, designers and technicians of Governments, project owners design units, construction enterprise, supervision companies, property companies, software vendors and system integrators.
特点 Features	图文并茂，突出要点，实操性强，中英文对照（中英文不一致时以中文为准），有较强的指导性、参考性、实用性和权威性。 Combination of words and photographs, highlighting key points, powerful practical functionality, Chinese and English versions（In case of discrepancy between Chinese and English, the Chinese text shall prevail）, strong guidance, reference, practical and authoritative values
不足 Inadequacies	由于大家均是利用业余时间、短时间编写，又采用了大量的国内、外的BIM资料，翻译的经验也不足，有不妥或不准确之处，请大家批评指正。 Incongruity and inaccuracy are unavoidable due to the limited compilation time as this was done during spare time, as well as abundant domestic and foreign BIM data and insufficient translation experience. Constructive feedback is welcome.

中国建筑设计研究院（集团）
院长助理、教授级高工
国务院政府特殊津贴专家
2013-7-1
Ouyang Dong
China Architecture Design & Research Institute (Group)
Assistant President, Professor-level Senior Engineer
State Council Expert with Special Allowance
JUL 1, 2013

目录 CONTENTS

1 BIM技术的现状和发展趋势
Current Status and Future Trend of BIM Technology

1.1 BIM技术总论 … 2
Overview of BIM Technology … 2

1.2 BIM技术的应用现状 … 9
Current Application of BIM Technology … 9

1.3 BIM技术的主要特点 … 37
Major Features of BIM Technology … 37

1.4 BIM技术未来的发展趋势 … 42
Future Trends of BIM Technology … 42

2 BIM技术应用的优势和劣势比较
Comparison of Application of the Advantages and Disadvantages of BIM Technology

2.1 二维设计软件与BIM设计软件的十四个比较 … 50
14 Top Differences Between 2D Design Software and BIM Design Utilities … 50

2.2 BIM技术的八大设计优势 … 64
Eight Advantages of BIM Technology … 64

2.3 BIM技术的八项绿色建筑分析 … 65
Eight Analysis of Green Building by Using BIM Technology … 65

3 BIM技术标准现状及编制规划
Current Status and Compiling Planning of BIM Technical Standards

3.1 国外与中国香港BIM技术标准现状 … 67
Current Status of BIM technical standards Abroad and in Hong Kong … 67

3.2 中国大陆地区BIM技术标准现状及编制规划 … 73
Current Status and Preparation Planning of BIM Technical Standards in the Chinese mainland … 73

3.3 BIM技术标准简述 … 78
Brief Introduction of BIM Technical Standards … 78

4 BIM技术在企业应用方面的推广
BIM Promotion in Enterprise Applications

4.1	BIM 设计技术发展模式	84
	Development Model of BIM Design Technology	84
4.2	中国建筑设计研究院BIM技术发展历程	86
	Development History of BIM in CAG	86
4.3	中国院在企业BIM方面推广的十一个步骤	87
	Eleven Promotion Steps of BIM in Enterprises by CAG	87
4.4	参加行业BIM专题讲演推广BIM技术	109
	Promoting BIM by Participating in Industry Special Lectures on BIM	109
4.5	BIM技术设计培训的八个步骤	111
	Eight Steps of BIM Technology Design Training	111
4.6	BIM技术设计协同	120
	BIM Technological Design Collaboration	120

5 BIM技术在设计项目方面的推广
BIM Promotion in Design Projects

5.1	BIM设计项目方面的八个推广步骤	132
	Eight Steps of Promoting BIM in Design Projects	132
5.2	中国院2010年BIM应用部分项目	141
	CAG's Implementation of BIM in 2010	141
5.3	中国院BIM应用代表案例——某金融商务中心	144
	Typical Case CAG using BIM Technology: A Financial Business Centre	144
5.4	中国院BIM应用经典案例——北京某信息产业基地	150
	Classic Case of CAG using BIM Technology: Beijing XX Information Industry Base Project	150

6 BIM技术设计应用软件汇总
Summary of BIM Technology in the Design of Application Software

- 6.1 BIM技术软件汇总之一（美国Autodesk公司） 156
  Summary of BIM Tool Applications 1（U.S.A Autodesk Inc.） 156
- 6.2 BIM技术软件汇总之二（美国Bentley公司） 158
  Summary of BIM Tool Applications 2（Bentley Systems U.S.A Ltd.） 158
- 6.3 BIM技术软件汇总之三（芬兰Progman公司） 160
  Summary of BIM Tool Applications 3（Finland Progman Co., Ltd.） 160
- 6.4 BIM技术软件汇总之四（北京理正软件公司） 162
  Summary of BIM Tool Applications 4（Beijing Leading Software Co., Ltd.） 162
- 6.5 BIM技术软件汇总之五（北京互联立方技术服务有限公司） 164
  Summary of BIM Tool Applications 5（Beijing isBIM Technical Services Co., Ltd.） 164
- 6.6 BIM技术软件汇总之六（北京鸿业同行科技公司） 166
  Summary of BIM Tool Applications 6（Beijing Hongye Tongxing Technology, Ltd.） 166

7 BIM技术应用遇到的问题和政府影响
Problems in BIM Technology Implementation and the Role of Government

- 7.1 BIM技术应用遇到的问题 169
  Problems in BIM Technology Implementation 169
- 7.2 需要政府、协会在推进BIM技术过程中解决的问题 171
  Problems to be Resolved in Promotion of BIM Technology by Government and Associations 171
- 7.3 政府的BIM技术政策 173
  Government BIM Technology Policies 173
- 7.4 BIM技术发展的思考 175
  Reflections on BIM Technological Development 175

附录
Appendix

1. 中国建筑设计研究院
   China Architecture Design & Research Group (CAG)
2. 美国Autodesk公司
   U.S.A Autodesk Inc.
3. 美国Bentley公司
   Bentley Systems U.S.A. Ltd.
4. 芬兰Progman公司
   Finland Progman Co., Ltd.
5. 北京理正软件公司
   Beijing Leading Software Co., Ltd.
6. 北京互联立方技术服务有限公司
   Beijing is BIM Technical Services Co., Ltd.
7. 北京鸿业同行科技公司
   Beijing Hongye Tongxing Technology Co., Ltd.

附光盘
Disk

1. 中国院BIM应用代表案例——某金融商务中心的"建筑、结构、机电动画"。
   Typical Case of CAG Using BIM Technology-"Animation of Architecture, Structure, MEP" of A Financial Business Center.
2. 中国院BIM应用经典案例——北京某信息产业基地项目"建筑综合动漫"。
   Classic Case of CAG Using BIM Technology-"Animation of Building Complex" of Beijing XX Information Industry Base Project.

本书重点摘要
Summary

1 BIM技术的现状和发展趋势
Current Status and Future Trend of BIM Technology

1.1 BIM技术总论
Overview of BIM Technology

建筑行业全生命周期（设计、施工、运营）的BIM技术革命
The BIM Revolution in the Full Lifecycle of the Construction Industry (Design, Construction and Operation)

BIM技术建筑行业（设计、施工、运营）技术的8个特点
Eight Features of BIM Technology in the Technology of the Construction Industry (Design, Construction and Operation)

	BIM技术建筑行业（设计、施工、运营）技术的8个特点 Eight Features of BIM Technology in the Technology of the Construction Industry (Design, Construction and Operation)		
1	可视化 Visualization	可视化 Visualization	可视化 Visualization
2	参数驱动 Parameter driven	专业综合 Professional integration	信息与分类标准 Information and classification standard
3	关联修改 Correlation modification	施工过程模拟 Construction process simulation	物资编码与管理 Material coding and management
4	任务划分与管理 Task partitioning and management	施工工艺模拟 Construction process simulation	远程与移动平台工作 Remote and mobile platform work
5	性能分析 Performance analysis	施工信息管理 Construction information management	基于BIM信息的设备管理 BIM information-based device management
6	协同设计 Collaborative design	基于BIM信息成本管理 BIM information-based cost management	基于BIM信息的租赁管理 BIM information-based lease management

（续表）

7	三维设计交付 3D design delivery	物资编码与管理 Material coding and management	基于BIM信息的安全管理 BIM information-based security management
8	远程与移动平台工作 Remote and mobile platform work	远程与移动平台工作 Remote and mobile platform work	基于BIM信息的自控系统 BIM information-based auto-control system

BIM技术能解决各方的主要关注点
Major focuses which can be settled by using BIM technology

序号 SN	各方名称 Name	BIM技术能解决各方的主要关注点 Major focuses which can be settled by using BIM technology				
		关注点1 Focus 1	关注点2 Focus 2	关注点3 Focus 3	关注点4 Focus 4	关注点5 Focus 5
1	政府方 Government				行业标准 Industry standard	技术进步 Technological improvement
2	建设方 Project Owner				工程进度 Project progress	风险控制 Risk control
3	设计方 Project Designer	工程质量 Engineering quality	成本控制 Cost control	工作效率 Work efficiency	设计收入 Design revenue	业务范围 Business scope
4	施工方 Project Contractor				工程进度 Project progress	修改返工 Modification and reworking
5	运营方 Project Operator				信息传递 Information transfer	维护便捷 Convenient maintenance

1.2 BIM技术的应用现状
Current Application of BIM Technology

- 美国使用BIM技术的现状
- 欧洲使用BIM技术的现状
- 亚洲使用BIM技术的现状
- 中国使用BIM技术的现状

BIM in the US
BIM in Europe
BIM in Asia
BIM in China

附表 Annexed tables

中国勘察设计协会主办《第一届"创新杯"BIM设计大赛（2010）》
The 1st "Innovation Cup" for BIM Design (2010) sponsored by China Exploration & Design Association

中国勘察设计协会主办《第二届"创新杯"BIM设计大赛（2011）》
The 2nd "Innovation Cup" for BIM Design (2011) sponsored by China Exploration & Design Association

中国勘察设计协会主办《第三届"创新杯"BIM设计大赛（2012）》
The 3rd "Innovation Cup" for BIM Design (2012) sponsored by China Exploration & Design Association

1.3 BIM技术的主要特点
Major Features of BIM Technology

- 特点之一：可视化设计
- 特点之二：参数化设计
- 特点之三：关联修改
- 特点之四：任务划分与管理
- 特点之五：性能分析
- 特点之六：协同设计
- 特点之七：三维设计交付
- 特点之八：远程与移动平台工作

Feature 1: Visualization design
Feature 2: Parametric design
Feature 3: Correlation modification
Feature 4: Task partitioning and management
Feature 5: Performance analysis
Feature 6: Collaborative design
Feature 7: 3D design delivery
Feature 8: Remote and mobile platform work

1.4 BIM技术未来的发展趋势
Future Trends of BIM Technology

- 趋势之一：国家发展目标与BIM未来技术发展相一致
- 趋势之二：未来BIM的发展整体架构图
- 趋势之三：BIM技术促进了决策流程和成本控制的优化
- 趋势之四：BIM技术应用的高价值体现
- 趋势之五：5D技术对项目成本、周期、质量的影响力
- 趋势之六：云计算对建筑产业发展的影响
- 趋势之七：BIM技术对智能建筑及数字城市的技术支撑
- 趋势之八：绿色可持续及装配式建筑设计

Trend 1: Consistency of national development goals and future BIM technical development

Trend 2: Overall framework of future BIM development

Trend 3: BIM technology optimizes decision-making flow and cost control

Trend 4: High value of BIM application

Trend 5: Influence of 5D technology on project cost, cycle and quality

Trend 6: Influence of cloud computing on the development of the construction industry

Trend 7: Technical support of BIM technology in smart buildings and digital cities

Trend 8: Green sustainability and fabricated architectural design

2 BIM技术应用的优势和劣势比较
Comparison of Application of the Advantages and Disadvantages of BIM Technology

2.1 二维设计软件与BIM设计软件的十四个比较
14 Top Differences Between 2D Design Software and BIM Design Utilities

- **对比之一：** 设计信息在整个设计过程中的传递关系

  Difference 01: Delivery of design information between each stage and discipline in project

  对比结果：二维设计软件的信息在不同工程阶段及不同专业间的传递有损失；而三维BIM设计软件可实现信息更有效传递！

  Conclusion: Information is easily lost when delivering information in 2D design software, between different stages and different disciplines, but 3D BIM software can deliver the information more efficiently!

• **对比之二**：设计工作量、设计过程的重心变化关系

Difference 02: Change of focus in terms of design workload and design process

对比结果：为保证设计质量，二维设计软件设计人员的大量时间花在协调和对图上，后期修改工作量大；而三维BIM设计软件设计工作量前移，重点在方案比选和技术优化，一旦模型关联关系建立修改便利！

Conclusion: To assure the design quality, the 2D designers spend much time in coordinating and comparing drawing, and will suffer heavy work load if any change in drawing after; however, the 3D BIM software can put the word in advance, and focus on the scheme choice and technology optimization. When the model is setup, it is easy to make change.

• **对比之三**：计算与绘图的融合修改关系

Difference 03: Integration of calculation and drafting

对比结果：二维设计软件的专业计算基本与绘图脱节；而三维BIM设计软件将计算与绘图融合，做到一处修改，处处更新。

Conclusion: Calculation and drafting are separated by 2D design software. They can be combined by BIM design software, enabling updates in real time.

• **对比之四**：二维图标与真实产品构件库的关系

Difference 04: Relationship between 2D symbols and real products

对比结果：二维设计软件CAD图块仅表达外形，相关数据缺失；而三维BIM设计软件可提供"数形合一"的构件族库！

Conclusion: AutoCAD 2D blocks created by 2D design software include only geometry without any technical data. With BIM design software, product databases with real products can contain both geometry and technical data.

• **对比之五**：设备数据与工作状态模拟的关系

Difference 05: Relationship between equipment data and project progress simulations

对比结果：二维设计软件无法进行设备工作状态模拟；而三维BIM设计软件通过设备构件族信息，可准确模拟设备的工作状态！

Conclusion: Project progress simulations are not supported by 2D design software, but they can be supported by BIM design software. This is made possible by product databases that contain complete technical data on real, commercially available products.

- **对比之六：** 平面、立面及剖面的对应关系

  Difference 06: The corresponding relationship between plane, elevation and section

  对比结果：二维设计软件的平面、立面和剖面相互对应是一个难题；而BIM设计软件可以实现模型和视图之间自动关联与更新！

  Conclusion: Associating plane, elevation and section is quite difficult using 2D design software. BIM design software makes it possible to associate and update them automatically.

- **对比之七：** 机电管线碰撞检测与综合的关系

  Difference 07: MEP pipeline collision detection and pipeline integration

  对比结果：二维设计软件的机电管线难以实现检测碰撞；而BIM设计软件通过机电碰撞检测与管线综合，实现高效、高质量的协同设计！

  Conclusion: Efficient collision detection between MEP pipelines cannot be achieved using 2D design software. BIM design software offers MEP pipeline collision detection and pipeline integration, making real-time coordination easy, efficient and high quality.

- **对比之八：** 机电管线与建筑结构的配合关系

  Difference 08: Coordination of MEP pipeline with construction

  对比结果：二维设计软件很容易产生碰撞，并浪费建筑有效空间；而BIM设计软件通过专业协同设计，减少碰撞，实现建筑空间的有效利用！

  Conclusion: With 2D design software, many collisions are not noticed before the construction stage and a great deal of space will be wasted. BIM design software enables coordination between different disciplines, making it possible to resolve collisions and use space more efficiently.

- **对比之九：** 机电管线预留洞口与土建预留洞口的配合关系

  Difference 09: Coordination of provisions for voids with construction

  对比结果：二维设计软件的机电管道与土建预留洞口经常冲突；而BIM设计软件通过专门的预留洞口功能，实现实时的预留洞口协同设计！

  Conclusion: Using 2D design software causes many differences and collisions between MEP pipelines and provisions for voids in the construction stage. Real-time coordination can be achieved using the Provision for Voids function provided by BIM design software.

- **对比之十**：设计信息与图形的融合关系

  Difference 10: Integration of design information and model

  对比结果：二维设计软件的CAD是"点、线、圆"组合和叠加，信息易丢失；而三维BIM设计软件是实体在软件中的虚拟模型，信息完整！

  Conclusion: With 2D design software, information is easily lost between project stages because drawings are created using basic CAD objects without any technical data. This will not happen with BIM design software because BIM models contain complete technical data.

- **对比之十一**：一维、二维、三维的信息融合关系

  Difference 11: Integration of information from 1D, 2D and 3D drawings

  对比结果：二维设计软件绘制一维图和二维图没有任何联系；而BIM设计软件使一维、二维、三维及信息完全融合成为可能！

  Conclusion: In 2D design software, there is no relationship between 1D drawings and 2D drawings; BIM design software allows information to be combined between 1D, 2D and 3D drawings.

- **对比之十二**：工程量及材料数量统计的准确性关系

  Difference 12: Exact quantities and volumes in construction and bills of materials

  对比结果：二维设计软件的工程量及材料量的统计准确度偏低；而BIM设计软件的统计信息是从虚拟模型中提取的，准确可信！

  Conclusion: It is impossible to count exact quantities and volumes in construction and bills of materials using 2D design software. BIM design software can do this automatically and reliably.

- **对比之十三**："数形合一"与设计质量的关系

  Difference 13: BIM model and design quality

  对比结果：二维设计软件受功能所限，很难在设计质量上有所突破；而BIM设计软件从协同设计上保证了设计质量，使设计结果更加可靠！

  Conclusion: The limitations of 2D design software make improving design quality practically impossible. Quality of design can be assured by coordination in BIM design software. Also the results will be more reliable.

- **对比之十四：** 设计信息的流动与传递关系
  Difference 14: Flow and delivery of design information
  对比结果：二维设计软件受平台和格式所限，无法实现设计结果和信息传递；而BIM设计软件保证了模型和信息的传递！
  Conclusion: The limitations of 2D design software make effective transfer of design results and information between disciplines impossible. With BIM design software, successful flow of design information can be assured.

2.2 BIM技术的八大设计优势
Eight Advantages of BIM Technology

序号 NO.	设计优势 Design Advantage	内容描述 Content Description
1	三维设计 3D Design	项目各部分拆分设计，便于特别复杂项目的方案设计，简单项目质量优化。 Break up project, convenient for the scheme design of complex project and quality optimization of easy project.
2	可视设计 Visualization Design	室内、室外可视化设计，便于业主决策，减少返工量。 Indoor or outdoor visualization design, convenient for owner to make decision, and reduce the rework.
3	协同设计 Collaboration Design	多个专业在同一平台上设计，实现了高效的协同设计。 Different disciplines work on the same platform; realize a high efficient collaboration design.
4	设计变更 Design alteration	一处修改，处处更新，计算与绘图的融合。 One change, and update in the whole project, to integrate the computation and drawing
5	碰撞检测 Collision test	通过机电专业的碰撞检测，解决机电管道碰撞。 By collision test in MEP, to eliminate the collision of pipelines
6	提高质量 Quality improvement	采用阶段协同设计，减少错漏碰缺，提高图纸质量。 Collaboration design in each phase can reduce the mistake and collision, to improve the drawing quality
7	自动统计 Automatic statistic	可自动统计工程量并生成材料表。 Calculate the quantity statistically, and create material list
8	节能设计 Energy saving design	支持整个项目绿色节能环保可持续发展。 Support the energy saving, environmental protection and sustainable development of the project.

2.3 BIM技术的八项绿色建筑分析
Eight Analysis of Green Building by Using BIM Technology

八项BIM技术绿色建筑分析

- 建筑能耗模拟分析
  simulation analysis of building energy consumption
- 遮阳与日照模拟分析
  simulation analysis of daylight and sunshade
- 室内、室外热辐射模拟分析
  Simulation analysis of indoor and outdoor heat radiation
- 室内舒适度模拟分析
  Simulation analysis of indoor comfort degree
- 通风效果模拟分析
  Simulation analysis of ventilation
- 建筑光环境分析
  Simulation analysis of luminous environment
- 气候仿真分析
  Simulation analysis of climate
- 地形仿真分析
  Simulation analysis of topography

3 BIM技术标准现状及编制规划
Current Status and Compiling Planning of BIM Technical Standards

3.1 国外与中国香港BIM技术标准现状
Current status of BIM technical standards Abroad and in Hong Kong

- 全球BIM发展
- 欧美及澳大利亚BIM标准
- 亚洲BIM标准

Global BIM development
BIM standards in Europe, America and Australia
BIM standards in Asia

3.2 中国大陆地区BIM技术标准现状及编制规划
Current Status and Preparation Planning of BIM Technical Standards in the Chinese mainland

- 中国BIM框架标准
- 中国BIM行业标准

Chinese BIM framework standard
Chinese BIM industrial standard

3.3 BIM技术标准简述
Brief Introduction of BIM Technical Standards

- BIM标准的层级划分
- 框架标准
- 行业标准
- 企业标准
- 项目标准
- BIM标准的编制基础
- 编制思路
- 软件平台
- 编制范围

Division levels of BIM standards
Framework standard
Industrial standard
Enterprise standard
Project standard
Basis for BIM standard preparation
Preparation ideas
Software platform
Scope of preparation

4 BIM技术在企业应用方面的推广
BIM Promotion in Enterprise Applications

4.1 BIM 设计技术发展模式
Development Model of BIM Design Technology

4.2 中国建筑设计研究院 BIM 技术发展历程
Development History of BIM in CAG

4.3 中国院在企业BIM方面推广的十一个步骤
Eleven Promotion Steps of BIM in Enterprises by CAG

- 第一步：宣传动员
- 第二步：技术研讨
- 第三步：组织架构
- 第四步：战略规划
- 第五步：专业培训
- 第六步：硬件建设
- 第七步：软件建设
- 第八步：课题研究
- 第九步：激励措施
- 第十步：BIM设计研究中心
- 第十一步：BIM科研楼

Step 1: Propaganda & mobilization
Step 2: Technical seminars
Step 3: Organizational structure
Step 4: Strategic planning
Step 5: Professional training
Step 6: Hardware configuration
Step 7: Software configuration
Step 8: Subject research
Step 9: Incentive policies
Step 10: BIM Design & Research Center
Step 11: BIM scientific research building

4.4 参加行业BIM专题讲演推广BIM技术
Promotion BIM by Participating in Industry Special Lectures on BIM

4.5 BIM技术设计培训的八个步骤
Eight Steps of BIM Technology Design Training

- 第一步：确定BIM培训目标
- 第二步：选择BIM培训对象
- 第三步：选择BIM培训机构
- 第四步：确定BIM培训教材
- 第五步：制定BIM培训计划
- 第六步：BIM培训过程管理
- 第七步：BIM培训考核认证
- 第八步：BIM培训归纳总结

Step 1: Identify BIM training goals
Step 2: Select BIM trainees
Step 3: Select BIM training institutions
Step 4: Select BIM training materials
Step 5: Develop BIM training plan
Step 6: BIM training process management
Step 7: BIM training assessment and certification
Step 8: BIM training summary and conclusion

4.6 BIM技术设计协同
BIM Technological Design Collaboration

- BIM多方协同
- BIM设计阶段的协同要点
- BIM多平台协同
- BIM与传统二维平台的协同要点
- BIM与参数化的协同要点
- BIM与绿色节能的协同要点
- BIM多专业协同
- BIM工作模式的要点
- BIM协同平台的技术实施方案

BIM multi-party collaboration
Key points for collaboration at BIM design stage
BIM multi-platform collaboration
Key points for BIM collaboration with conventional 2D platforms
Key points for BIM collaboration with parameters
Key points for BIM collaboration with green energy-saving
BIM Multi-disciplinary collaboration
Key points of BIM working mode
Technical implementation plans for BIM collaboration platforms

5 BIM技术在设计项目方面的推广
BIM Promotion in Design Projects

5.1 BIM设计项目方面的八个推广步骤
Eight Steps of Promoting BIM in Design Projects

- 第一步：BIM项目目标
- 第二步：BIM人员配置
- 第三步：BIM硬件及软件配置
- 第四步：BIM计划和措施
- 第五步：BIM技术培训
- 第六步：BIM过程质量管理
- 第七步：方案、初设、施工图BIM技术应用点
- 第八步：BIM成果总结

Step 1: BIM project goals
Step 2: BIM staffing
Step 3: BIM hardware and software configuration
Step 4: BIM plan and measures
Step 5: BIM technical training
Step 6: BIM process quality management
Step 7: BIM technological focus on schematic design, preliminary design and working drawing design
Step 8: summary of BIM outcome

5.2 中国院2010年BIM应用部分项目
CAG's implementation of BIM in 2010

- 中国院 BIM 设计实践范围
- 中国院 BIM设计案例实践线路
- 中国院BIM设计应用部分项目

CAG's practice scope of BIM design
CAG's practice path of BIM design
CAG's projects using BIM design

5.3 中国院BIM应用代表案例——某金融商务中心
Typical Case CAG using BIM Technology: A Financial Business Centre

- 某金融商务中心项目概况和获奖情况
- 某金融商务中心五个BIM技术创新点
- 创新点之一：建筑施工图100%采用BIM平台输出
- 创新点之二：Revit模型导入结构计算软件
- 创新点之三：基于机械制造业技术的局部细化设计
- 创新点之四：模板图及配筋图的BIM绘图技术
- 创新点之五：基于产品实际技术参数的设计技术

Project Overview and Awards of A Financial Business Centre
Five innovative highlights of BIM technology for a financial business centre
Innovative highlight 1: architectural working drawings fully adopt BIM platform for output
Innovative highlight 2: structural computation software is imported into the revit model
Innovative highlight 3: local detailed design based on machine manufacturing technology
Innovative highlight 4: BIM drawing technology in template and reinforcement drawings
Innovative highlight 5: design technology based on actual technical parameters

5.4 中国院BIM应用经典案例——北京某信息产业基地
Classic Case of CAG using BIM Technology: Beijing XX Information Industry Base Project

- 北京某信息产业基地项目概况及获奖情况
- 某国际信息港二期五个**BIM**精细化设计
- 精细化设计之一——多方案比选
- 精细化设计之二——管线综合
- 精细化设计之三——三维的思考设计
- 精细化设计之四——绿色仿真模拟计算
- 精细化设计之五——能耗及舒适度分析

Project Overview and Awards of Beijing XX Information Industry Base Project five highlights of BIM-based refined design on international information port (phase Ⅱ)
Refined design highlight 1: multi-scheme comparison
Refined design highlight 2: pipeline integration
Refined design highlight 3: design with 3D methodology
Refined design highlight 4: green simulation calcalation
Refined design highlight 5: energy consumption and comfort analysis

6 BIM技术设计应用软件汇总
Summary of BIM Technology In the Design of Application Software

6.1 BIM技术软件汇总之一（美国Autodesk公司）
Summary of BIM Tool Applications 1（U.S.A Autodesk Inc.）

6.2 BIM技术软件汇总之二（美国Bentley公司）
Summary of BIM Tool Applications 2（Bentley Systems U.S.A Ltd.）

6.3 BIM技术软件汇总之三（芬兰Progman公司）
Summary of BIM Tool Applications 3（Finland Progman Co. Ltd.）

6.4 BIM技术软件汇总之四（北京理正软件公司）
Summary of BIM Tool Applications 4（Beijing Leading Software Co. Ltd.）

6.5 BIM技术软件汇总之五（北京互联立方技术服务有限公司）
Summary of BIM Tool Applications 5（Beijing isBIM Technical Services Co., Ltd.）

6.6 BIM技术软件汇总之六（北京鸿业同行科技公司）
Summary of BIM Tool Applications 6（Beijing Hongye Tongxing Technology, Ltd.）

7 BIM技术应用遇到的问题和政府影响
Problems in BIM Technology Implemention and the Role of Government

7.1 BIM技术应用遇到的问题
Problems in BIM Technology Implementation

- BIM文件的法律责任问题
- BIM技术规范和标准问题
- BIM的应用和交付深度问题
- BIM技术软件的不完善问题
- BIM技术文件建设档案馆存档问题
- BIM技术各种成本过大问题

Legal liability of BIM files
BIM technical specifications and standards
BIM application and delivery depth
BIM Technology software imperfections
Archiving of BIM technical documents and construction archives
High cost of BIM technology

7.2 需要政府、协会在推进BIM技术过程中解决的问题
Problems to be Resolved in Promotion of BIM Technology by Government and Associations

- 进行BIM技术的宣传推广
- 制定BIM技术标准和技术措施
- 制定BIM技术管理办法和奖励政策
- 出台"BIM技术报批制度"
- 组织协调相关政府部门解决BIM审批及存档问题
- 制定"BIM技术设计收费指导办法"

Promotion of BIM technology
Formulate BIM technical standards and measures
Formulate BIM technical administration measures and incentive policies
Issue BIM technology approval system
Settle BIM approval and archiving by organizing and coordinating relevant government departments
Formulate guidance measures of BIM technical design fees

7.3 政府的BIM技术的政策
Government BIM Technology Policies

- 美国政府BIM技术政策
- 新加坡政府BIM技术政策
- 中国政府BIM技术政策

The US government's BIM technology policies
The Singapore government's BIM technology policies
The Chinese government's BIM technology policies

7.4 BIM技术发展的思考
Reflections on BIM Technological Development

- BIM技术发展的思考之一：适合设计企业发展的BIM应用格局
- BIM技术发展的思考之二：基于设计行业的BIM发展
- BIM技术发展的思考之三：BIM设计的六个阶段

Reflection 1: BIM application pattern suitable for design firms' development
Reflection 2: BIM development based on design industry
Reflection 3: Six stages of BIM design

附录 Appendix

1. 中国建筑设计研究院
   China Architecture Design & Research Group (CAG)
2. 美国Autodesk公司
   U.S.A Autodesk Inc.
3. 美国Bentley公司
   Bentley Systems (USA) Co., Ltd.
4. 芬兰Progman公司
   Finland Progman Co., Ltd.
5. 北京理正软件公司
   Beijing Leading Software Co., Ltd.
6. 北京互联立方技术服务有限公司
   Beijing isBIM Technical Services Co., Ltd.
7. 北京鸿业同业科技公司
   Beijing Hongye Tongxing Technology Co., Ltd.

附光盘 Disk

1 中国院BIM应用代表案例——某金融商务中心的"建筑、结构、机电动漫";
Typical Case of CAG Using BIM Technology- "Animation of Architecture, Structure, MEP" of A Finamcial Business Center.

2 中国院BIM应用经典案例——北京某信息产业基地项目"建筑综合动漫"。
Classic Case of CAG Using BIM Technology- "Animation of Building Complex" of Beijing XX Information Industry Base Project.

BIM技术的现状和发展趋势

Current Status and Future Trend of BIM Technology

社会与经济的快速发展,给建筑行业既带来高速发展的机遇,也带来不断的挑战。项目日益复杂,业主要求越来越高,市场竞争激烈,这都要求建筑设计企业进行变革突破,从而提升核心竞争力,迎接挑战。在这一背景下,近年迅速发展起来的建筑信息模型(Building Information Modeling,即BIM)这一理念被广泛认可为未来设计行业的发展趋势。在美国,大部分建筑设计企业已经开始采用BIM技术进行设计;在中国,部分建筑设计单位也已经开始在应用BIM技术方面有所突破,取得了一定的成果。

Rapid social and economic development has brought many opportunities for the construction industry to develop, as well as continuous challenges. Projects are becoming more complex, clients are more demanding, and the market is becoming more competitive, all requiring architectural firms to change and make their own breakthroughs to improve their core competitiveness in confronting challenges. With this background, Building Information Modeling (BIM), a quickly emerging concept in recent years, is widely recognized as the future trend in design. In the United States, a large number of architectural firms have started using BIM technology for their designs; in China, some architectural firms have also used BIM and made some achievements.

1.1 BIM技术总论
Overview of BIM Technology

BIM技术——建筑行业(设计——施工——运营)全生命周期的技术革命
The BIM Technology Revolution in the Full Lifecycle of the Construction Industry (Design, Construction and Operation)

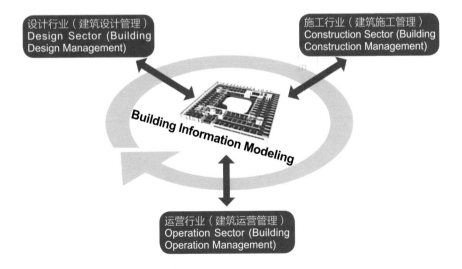

BIM是建筑业的商业信息管理平台
BIM serves as a business information management platform for the construction industry

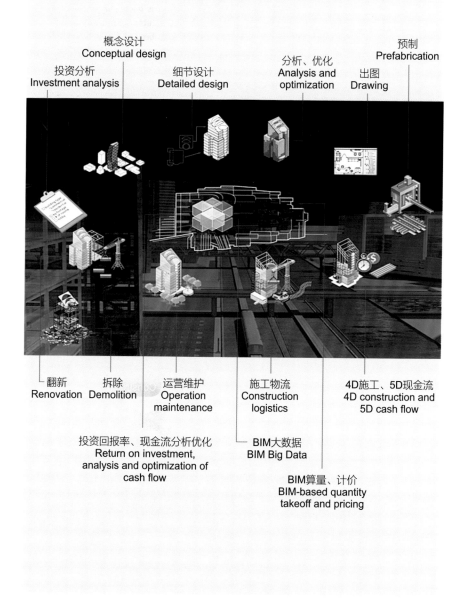

1. BIM技术建筑行业（设计、施工、运营）变革的8个特征
Eight Features of BIM Technology in Revolutionizing the Construction Industry (Design, Construction and Operation)

序号 SN	特征点 Feature Points	建筑设计变革的特征 Features of Changing in the Building Design	建筑施工变革的特征 Features of Changing in the Building Construction	建筑运营变革的特征 Features of Changing in the Building Operation
1	国家目标 The National Goal	住房和城乡建设部《2011－2015年建筑业信息化发展纲要》明确提出：加快推广BIM、协同设计、移动通信、无线射频、虚拟现实、4D项目管理等技术在勘察设计、施工和工程项目管理中的应用，改进传统的生产与管理模式，提升企业的生产效率和管理水平。将推动BIM技术发展在建筑行业（设计、施工、运营）应用列为企业的重点发展目标。 *The Guidelines for Information Development of the Construction Industry 2011-2015* formulated by the Chinese Ministry of Housing and Urban-Rural Development expressly states: to accelerate the promotion of BIM, collaborative design, mobile communication, radio frequency, virtual reality, and 4D project management in engineering survey, design, construction and project management; improve traditional production and management modes, and enhance business productivity and management level. Encourage construction companies to focus on BIM technology as their developing target in the design, construction and operation processes.		
2	绿色节能 Green Energy-saving	根据国家的节能减排政策，设计节能环保的绿色建筑将逐渐成为对建筑设计行业的必然要求。 According to China's energy-saving and emission reduction policies, designing green buildings that are energy efficient and environmentally friendly will become a necessity in the design sector.	根据国家"四节一环保"政策，建造节能、绿色、生态建筑将成为建筑施工行业的发展趋势。 In line with the national policy of "Four Savings and One Environmental Protection", constructing energy-saving, green and ecological buildings will become a future trend in the construction sector.	美国绿色建筑协会的"LEED绿色建筑认证体系"、中国住房和城乡建设部的"绿色三星建筑认证标准"。 The LEED Green Building Rating System developed by the U.S. Green Building Council, and the Three Star Building Certification Standard developed by the Chinese Ministry of Housing and Urban-Rural Development.

（续表1）

序号 SN	特征点 Feature Points	建筑设计变革的特征 Features of Changing in the Building Design	建筑施工变革的特征 Features of Changing in the Building Construction	建筑运营变革的特征 Features of Changing in the Building Operation
3	行业发展 Industrial Development	BIM技术不仅是一次建筑设计革命，而是整个建筑产业的革命。BIM技术应用将成为建筑设计行业基本要求的趋势正逐渐明朗。 BIM technology is not only revolution in architectural design, but in the entire construction industry. BIM application will gradually become the basic requirement of the design sector.	根据美国相关数据，建筑业存在57%的浪费，存在较大的提升空间。BIM技术作为整个产业的革命，必将对施工管理、施工质量、建造成本等产生巨大的影响。 Based on data from the U.S., there is up to 57% waste in the construction sector, which means there's much room for improvement. As it revolutionizes the entire industry, BIM technology will fundamentally influence construction management, quality and cost control.	利用竣工模型提供的信息，高效准确地管理和维护设施运行，以更低的运维成本最大限度发挥设施效用。 Use the information presented by the completed model to accurately and efficiently manage and maintain operation of facilities, and maximize facility use with minimum operation cost and maintenance.
4	市场竞争 Market Competition	设计市场竞争日趋激烈，面临僧多粥少的局面。业主的要求越来越高，设计技术也必须与时俱进，才能提高企业的核心竞争力。 The design market becomes increasingly competitive, and clients are more demanding. Under these circumstances, design technology should also adapt with the times to improve core business competitiveness.	当前施工市场企业数量众多，竞争激烈，施工效率低下，施工技术含金量不高，通过BIM实现施工市场、人才资源的优化重组，实现良性竞争。 Currently, the construction market has too many competitors with low efficiency and technical value. By adopting BIM technology, the construction market and human resources can be optimized and create healthy competition.	面对用户的不同需求，物业公司缺少先进的运营维护手段，管理水平参差不齐，竞争激烈。通过BIM技术提供系统的运营维护解决方案。 Facing different requirements of users, property management companies usually lack advanced operation and maintenance methods. Their management levels are varied and competition is fierce. BIM technology can provide operation and maintenance solutions in a systematic way.

（续表2）

序号 SN	特征点 Feature Points	建筑设计变革的特征 Features of Changing in the Building Design	建筑施工变革的特征 Features of Changing in the Building Construction	建筑运营变革的特征 Features of Changing in the Building Operation
5	项目复杂性 Project Complexity	多功能、复杂、独特性的建筑越来越多，传统的设计方式已难以满足，必须采用BIM技术才能实现。With an increasing number of multifunctional, complex and unique buildings, traditional design methods cannot meet the needs of modern requirements. Application of BIM technology becomes a necessity.	通过BIM技术解决多功能复杂项目的施工工艺、项目管理等问题，提高施工质量，缩短工期，节约投资。BIM technology can solve the construction process, project management and other problems of multifunctional and complex projects, improve construction quality, shortens construction time and save money.	利用BIM技术，辅助了人工维护，实现高效运维，智能化管控，同时可为客户提供更多的增值服务。BIM technology can replace manual maintenance, realize highly efficient operation and maintenance, and smart management and control, as well as offer more value-added services for clients.
6	信息化发展 Information Development	BIM是一个可资源共享的信息平台，为参与方在各阶段提供决策依据。BIM技术将是设计行业投资未来的最佳途径之一。BIM is an information platform with shared resources that provides a decision-making basis for participants in all stage and will become one of the best ways for future investment in the design sector.	BIM可实现数据信息从设计有效传输到施工阶段，提升施工企业核心竞争力。BIM can realize effective transmission of data information from design to construction stages and improve contractors' core competiveness.	根据从设计、竣工三维模型传递来的信息（设备产品信息、参数等）作为运营维护的依据，可以节约大量的时间、人力、成本。Based on information (device product information, parameters, etc.) transferred for design and completed 3D model, it can save a lot of time, labor and cost in project operation and maintenance.

（续表3）

序号 SN	特征点 Feature Points	建筑设计变革的特征 Features of Changing in the Building Design	建筑施工变革的特征 Features of Changing in the Building Construction	建筑运营变革的特征 Features of Changing in the Building Operation
7	精细化 Refinement	由于对建筑细节要求、成本控制越来越高，只有通过BIM技术才能达到精细化设计的要求。 As requirements on building details and cost control increase, only BIM technology can achieve the requirements of refined design.	BIM改变了传统的粗放式施工组织管理模式，实现精细化管理，显著增强施工企业市场竞争力。 BIM transforms the traditional extensive construction organization and management mode into refined management, which can significantly enhance the market competitiveness of contractors.	通过BIM三维运维模型实现了运营维护的预见性、快速性、准确性。 The 3D completion model helps to realize predictability, speedy and accuracy of operation and maintenance.
8	BIM人才 BIM Talents	BIM设计打破原有单一技能的设计团队模式，转变为多元人才结构的复合型团队模式。 BIM technology can transform the original design team mode of a single skill into a complex team mode with diverse talents.	需要培养适应BIM新技术、新的施工组织模式的施工技术人才。 Training is needed to develop technical talents for construction to adapt to BIM new technology and new construction organizing mode.	通过BIM技术，可减少运营维护人员总数，但对运营维护人员提出了较高的技术要求。 BIM technology can reduce the total number of operation and maintenance staff, and have higher technical requirements.

2. BIM技术建筑行业（设计、施工、运营）技术的8个特点
Eight Features of BIM Technology in the Technology of the Construction Industry (Design, Construction and Operation)

序号 SN	建筑设计技术的8个特点 Eight Features of the Building Design Technology	建筑施工技术的8个特点 Eight Features of the Building Construction Technology	建筑运营技术的8个特点 Eight Features of the Building Operation Technology
1	可视化 Visualization	可视化 Visualization	可视化 Visualization
2	参数驱动 Parameter driven	专业综合 Professional integration	信息与分类标准 Information and classification standard
3	关联修改 Correlation modification	施工过程模拟 Construction process simulation	物资编码与管理 Material coding and management
4	任务划分与管理 Task partitioning and management	施工工艺模拟 Construction process simulation	远程与移动平台工作 Remote and mobile platform work
5	性能分析 Performance analysis	施工信息管理 Construction information management	基于BIM信息的设备管理 BIM information-based device management
6	协同设计 Collaborative design	基于BIM信息成本管理 BIM information-based cost management	基于BIM信息的租赁管理 BIM information-based lease management
7	三维设计交付 3D design delivery	物资编码与管理 Material coding and management	基于BIM信息的安全管理 BIM information-based security management
8	远程与移动平台工作 Remote and mobile platform work	远程与移动平台工作 Remote and mobile platform work	基于BIM信息的自控系统 BIM information-based auto-control system

3. BIM技术能解决各方的主要关注点
Major focuses which can be settled by using BIM technology

序号 SN	各方名称 Name	BIM技术能解决各方的主要关注点 Major focuses which can be settled by using BIM technology				
		关注点1 Focus 1	关注点2 Focus 2	关注点3 Focus 3	关注点4 Focus 4	关注点5 Focus 5
1	政府方 Government	工程质量 Engineering quality	成本控制 Cost control	工作效率 Work efficiency	行业标准 Industrial standard	技术进步 Technological improvement
2	建设方 Project Owner				工程进度 Project progress	风险控制 Risk control
3	设计方 Project Designer				设计收入 Design revenue	业务范围 Business scope
4	施工方 Project Contractor				工程进度 Project progress	修改返工 Modification and rework
5	运营方 Project Operator				信息传递 Information transfer	维护便捷 Convenient maintenance

1.2 BIM技术的应用现状
Current Application of BIM Technology

1. 美国使用BIM技术的现状
BIM application in the US

1）美国采用BIM的项目数量增长迅速，美国62%以上设计单位采用BIM设计技术，美国政府负责建设的项目，要求全部使用BIM技术。

1) The number of projects applying BIM in the US has significantly increased, with 62% of US design firms using BIM technology in their work, and BIM is a necessity for all projects related to the US government.

2）根据McGraw-Hill Construction公司发布《北美BIM商业价值评估报告（2007—2012）》中的数据：整个北美建筑行业BIM技术的应用迅速普及，2007—2012年BIM应用普及率的发展如下图所示。

2) According to *The Business Value of BIM in North America* (2007-2012) published by McGraw-Hill Construction, from 2007 to 2012, BIM was quickly and largely adopted by the construction industry in North America as shown in the figures below:

2007 **28%**   2009 **49%**   2012 **71%**

2012年，北美大型建筑企业的BIM应用率达到91%，中型建筑企业为86%，小型建筑企业为49%；根据美国BDC公司2012年7月发布的报告：美国AEC行业的300强企业均已应用BIM技术，并获得了较高的回报。

In 2012, the application rate of BIM in large construction companies in North America was as high as 91%, medium-sized construction firms at 86%, and small-sized ones at 49%. According to a report issued by BDC in July 2012, the top 300 firms in architectural, engineering and contracting fields (AEC) in the US achieved high returns on investment from using BIM.

3）美国BIM应用项目

3) Projects applying BIM in the US

BIM技术应用案例——美国某大厦
BIM application case: A building in the USA

2. 欧洲使用BIM技术的现状
BIM application in Europe

1）英国：2011年5月，政府内阁办公室发布2011—2016《政府建筑业发展战略》。根据2012年2月英国建设行业网络(CSN)发表的调查报告：截至2012年1月，英国AEC企业BIM的使用率已达到57%，同比增幅显著。

2）德国：根据美国McGraw-Hill Construction公司2010年的调查报告：2010年，德国的BIM应用率为36%，且其国产的BIM软件占主导地位。

3）法国：2010年法国BIM的应用率为38%。

4）BIM应用案例

1) The United Kingdom: In May 2011, the UK Cabinet Office published *UK Government Construction Strategy* 2011-2016. Data from the survey report published by UK Construction Skills Network (CSN) in February 2012 suggests that as of January 2012, the application of BIM in architectural, engineering and contracting firms (AEC) in the UK reached 57%, with significant year-on-year growth.

2) Germany: According to the 2010 survey report by McGraw-Hill Construction, BIM application in Germany that year was 36%, and its domestic-made BIM software was predominantly used.

3) France: BIM application in France in 2010 was 38%.

4) BIM application case

BIM技术应用案例——英国某大厦
BIM application case: A building in the UK

3. 亚洲使用BIM技术的现状
BIM application in Asia

1）新加坡：政府对信息化工作高度重视，建立了政府建筑监管审批的电子政务平台CORENET；设立国家BIM基金，对实施BIM的企业在初期给予现金补偿；提供免费BIM知识技能培训。

2）韩国：建立了一套从国家、行业部门到企业或项目级的比较完整的标准/指南体系，如：国家级的《国家BIM指南》(National BIM Guide)。

1) Singapore: the Singapore government pays close attention to information construction and has set up an E-government platform called CORENET to supervise and approve government construction projects. In addition, the national BIM Fund was established to help firms to adopt BIM technology into their work processes with cash compensation in the early stages. Training programs on BIM knowledge and skills were also provided free of charge.

2) Korea: Complete standard/guide systems for national, industrial, corporate and project levels were built in Korea such as the *National BIM Guide*.

4. 中国使用BIM技术的现状
BIM application in China

1）中国大陆：住房和城乡建设部发布了《2011-2015年建筑业信息化发展纲要》，2011年清华大学BIM标准研究课题组发布了《中国建筑信息模型标准框架研究（CBIMS）》。2012年中国勘察设计协会举办的第三届"创新杯"BIM大赛，参赛单位和项目数量、评委单位和评委人数、项目质量都有大幅提升。

2）香港：香港发展局和房屋署从2006年开始启动了一批BIM试点项目，并陆续发布了BIM标准、用户指南、构件资料库创建规范等设计指引和参考文件；2009年成立了香港BIM学会（HKIBIM）；2012年房屋署开始将BIM应用延伸到建筑施工领域。

1) The Chinese mainland: *Guidelines for Information Development of the Construction Industry 2011-2015* was formulated by the Chinese Ministry of Housing and Urban-Rural Development. In 2011, *Research of Chinese Building Information Modeling Standard Framework (CBIMS)* was published by Tsinghua BIM Research Group. In 2012, the 3rd "Innovation Cup" for BIM design was sponsored by China Exploration & Design Association (CEDA), where the number of participants, projects and judges, and project quality significantly increased.

2) Hong Kong: As early as 2006, the Development Bureau and Housing Department of Hong Kong started a series of BIM pilot projects, and have successively published design guidelines and reference documents on BIM standards, user guides and a component database. In 2009, the Hong Kong Institute of Building Information Modeling (HKIBIM) was established, and in 2012, the Housing Department started to extend the application of BIM to the construction field.

3）BIM应用案例　　　　　　　　　　3) BIM application cases

BIM技术应用案例——香港某火车站
BIM application case: A train station in Hong Kong

BIM技术应用案例——北京某瞭望塔
BIM application case: An observation tower in Beijing

BIM技术应用案例——上海某集团总部大楼
BIM application case: A group's headquarters building in Shanghai

BIM技术应用案例——上海中心
BIM application case: Shanghai Tower

北京"中国尊"项目
China Zun in Beijing

附表

中国勘察设计协会主办《第一届"创新杯"BIM设计大赛（2010）》

中国勘察设计协会主办《第二届"创新杯"BIM设计大赛（2011）》

中国勘察设计协会主办《第三届"创新杯"BIM设计大赛（2012）》

Annexed tables

The 1st "Innovation Cup" for BIM Design Competition (2010) sponsored by China Exploration & Design Association

The 2nd "Innovation Cup" for BIM Design Competition (2011) sponsored by China Exploration & Design Association

The 3rd "Innovation Cup" for BIM Design (2012) Competition sponsored by China Exploration & Design Association

1. 中国勘察设计协会《第一届"创新杯"BIM设计大赛（2010）》获奖名单

China Exploration & Design Association 〈 The 1st "Innovation Cup" for BIM Design Competition (2010) 〉 Winner List

《第一届"创新杯"BIM设计大赛（2010）》竞赛情况 Summary of The 1st "Innovation Cup" for BIM Design Competition (2010)			
竞赛时间 Competition Time	2010年8月 August, 2010	颁奖地点 Award prizes location	上海 Shanghai
参赛单位数量 Number of Participants	46个 46	参赛项目数量 Number of projects	147个 147
获奖单位数量 Number of winners	16个 16	奖项数量 Amount of awards	50个 50

中国勘察设计协会《第一届"创新杯"BIM设计大赛（2010）》获奖名单

China Exploration & Design Association〈The 1st "Innovation Cup" for BIM Design Competition (2010)〉Winner List

项目分类 Project classification		奖项 Award	单位名称 Organization name	项目名称 Project name
1	最佳BIM建筑设计奖 Best BIM for architectural design	一等奖 First Prize	现代集团华东建筑设计研究院有限公司 Xiandai Group East China Architectural Design & Research Institute Co., Ltd	世博文化中心 Expo Cultural Center
2			中国建筑设计研究院 China Architecture Design & Research Group	敦煌莫高窟游客中心 The Dunhuang Mogao Tourists Center
3			CCDI中建国际设计顾问有限公司 CCDI Group	世博国家电力馆 Expo State Grid Pavilion
4		二等奖 Second Prize	上海现代建筑设计（集团）有限公司 Shanghai Xian Dai Architectural Design （Group）Co., Ltd.	世博上汽通用企业馆 Expo SAIC-GM Pavilion
5			上海现代建筑设计（集团）有限公司 Shanghai Xian Dai Architectural Design （Group）Co., Ltd.	湖州喜来登温泉度假酒店 Huzhou Sheraton Resort & Spa
6			现代集团华东建筑设计研究院有限公司 Xiandai Group East China Architectural Design & Research Institute Co., Ltd.	世博上海案例馆 Expo Shanghai Pavilion
7			上海现代建筑设计（集团）有限公司 Shanghai Xian Dai Architectural Design （Group）Co., Ltd.	世博奥地利馆 Expo Austrian Pavilion
8			CCDI中建国际设计顾问有限公司 CCDI Group	杭州奥体中心体育馆 Hangzhou Olympic Center Gymnasium

(续表1)

项目分类 Project classification		奖项 Award	单位名称 Organization name	项目名称 Project name
9	最佳BIM建筑设计奖 Best BIM for architectural design	三等奖 Third Prize	CCDI中建国际设计顾问有限公司 CCDI Group	天津团泊湖网球中心 Tianjin Tuanbo Lake Tennis Center
10			CCDI中建国际设计顾问有限公司 CCDI Group	天津团泊湖新城综合体育馆 Tianjin Tuanbo Lake (Xincheng) Gymnasium
11			广东省建筑设计研究院 Guangdong Architectural Design and Research Institute	中山古镇灯都商厦 Zhongshan Old Town Light Commercial Building
12			现代设计集团上海建筑设计研究院有限公司 Xiandai Group Shanghai Architectural Design & Research Institute Co., Ltd	合肥滨湖国际会展中心 Hefei Bin Lake International Convention and Exhibition Center
13			CCDI中建国际设计顾问有限公司 CCDI Group	哈尔滨西客站 Harbin West Railway Station
14			中国建筑设计研究院 China Architecture Design & Research Group	徐州建筑职业技术学院图书馆 Xuzhou Institute Architectural Technology Library
15			现代设计集团华东建筑设计研究院有限公司 Xiandai Group East China Architectural Design & Research Institute Co. Ltd	申都大厦改建工程 Shanghai Shendu Renovation Project
16			深圳华森建筑与工程设计顾问有限公司 Huasen Architectural & Engineering Designing Consultant Ltd.	广州从化温泉养生谷商务会议Ⅲ区D型贵宾院落式酒店 Guangzhou Conghua Health Valley Spa Hotel

（续表2）

项目分类 Project classification	奖项 Award	单位名称 Organization name	项目名称 Project name
17	一等奖 First Prize	现代设计集团华东建筑设计研究院有限公司 Xiandai Group East China Architectural Design & Research Institute Co., Ltd.	世博文化中心 Expo Cultural Center
18		CCDI中建国际设计顾问有限公司 CCDI Group	世博国家电力馆 Expo State Grid Pavilion
19	最佳BIM工程设计奖 Best BIM for engineering design	现代设计集团华东建筑设计研究院有限公司 Xiandai Group East China Architectural Design & Research Institute Co. ,Ltd.	世博上海案例馆 Expo Shanghai Pavilions
20		CCDI中建国际设计顾问有限公司 CCDI Group	杭州奥体中心体育场 Hangzhou Olympic Center Gymnasium
21	二等奖 Second Prize	CCDI中建国际设计顾问有限公司 CCDI Group	天津团泊湖新城综合体育馆 Tianjin Tuanbo Lake (Xincheng) Gymnasium
22		CCDI中建国际设计顾问有限公司 CCDI Group	天津团泊湖网球中心 Tianjin Tuanbo Lake Tennis Center
23		机械工业第六设计研究院 SIPPR Engineering Group Co., Ltd.	浙江中烟工业公司杭州制造部"十一五"易地技术改造项目 China Tobacco Hangzhou Manufacturing Department "the 11th Five-Year Plan" Technological Transformation Projects

(续表3)

项目分类 Project classification	奖项 Award	单位名称 Organization name	项目名称 Project name
24	最佳BIM工程设计奖 Best BIM for engineering design	三等奖 Third Prize	华通设计顾问工程有限公司 Walton Design Consultant Engineering Co., Ltd.	世茂工三广场 The Shanghai Shimao Square
25			昆明市建筑设计研究院有限责任公司 Kunming Architectural Design & Research Institute Co., Ltd.	云南省第一人民医院住院综合楼 Yunnan First People's Hospital Building
26			东风设计研究院有限公司 Dongfeng Design Institute Group	某汽车工厂联合厂房 Automobile Factory Integrated Plant
27			上海现代建筑设计（集团）有限公司 Shanghai Xian Dai Architectural Design（Group）Co.,Ltd.	世博上汽通用企业馆 Expo SAIC-GM Pavilions
28			CCDI中建国际设计顾问有限公司 CCDI Group	宜昌万达广场 Yichang Wanda Square
29			上海现代建筑设计（集团）有限公司 Shanghai Xian Dai Architectural Design（Group）Co.,Ltd.	湖州喜来登温泉度假酒店 Huzhou Sheraton Resort & Spa
30	最佳BIM协同设计奖 Best BIM for collaboration design	一等奖 First Prize	现代集团华东建筑设计研究院有限公司 Xiandai Group East China Architectural Design & Research Institute Co., Ltd.	世博文化中心 Expo Cultural Center

（续表4）

项目分类 Project classification	奖项 Award	单位名称 Organization name	项目名称 Project name
31	二等奖 Second Prize	机械工业第六设计研究院 SIPPR Engineering Group Co., Ltd.	浙江中烟工业公司杭州制造部"十一五"易地技术改造项目 China Tobacco Hangzhou manufacturing department "11-5" technological transformation projects
32		CCDI中建国际设计顾问有限公司 CCDI Group	世博国家电力馆 Expo State Grid Pavilion
33 最佳BIM协同设计奖 Best BIM for collaboration design	三等奖 Third Prize	CCDI中建国际设计顾问有限公司 CCDI Group	杭州奥体中心主体育场 Hangzhou Olympic Sports Center Gymnasium
34		上海现代建筑设计（集团）有限公司 Shanghai XIANDAI Architectural Design (Group) Co. Ltd.	世博奥地利馆 Expo Austrian Pavilion
35		CCDI中建国际设计顾问有限公司 CCDI Group	哈尔滨西客站 Harbin West Railway Station
36 最佳BIM绿色分析应用奖 Best BIM for sustainable design	一等奖 First Prize	现代设计集团华东建筑设计研究院有限公司 Xiandai Group East China Architectural Design & Research Institute Co., Ltd.	世博文化中心 Expo Cultural Center
37	二等奖 Second Prize	CCDI中建国际设计顾问有限公司 CCDI Group	世博园国家电力馆 Expo State Grid Pavilion
38		CCDI中建国际设计顾问有限公司 CCDI Group	大梅沙万科中心 Dameisha Vanke Center

（续表5）

	项目分类 Project classification	奖项 Award	单位名称 Organization name	项目名称 Project name
39	最佳BIM绿色分析应用奖 Best BIM for sustainable design	三等奖 Third Prize	CCDI中建国际设计顾问有限公司 CCDI Group	凤凰岭绿色建筑节能项目 The Phoenix Ridge Green Building Energy-saving Project
40			现代设计集团华东建筑设计研究院有限公司 Xiandai Group East China Architectural Design & Research Institute Co., Ltd.	申都大厦改建工程 Shanghai Shendu Renovation Project
41			现代设计集团华东建筑设计研究院有限公司 Xiandai Group East China Architectural Design & Research Institute Co., Ltd.	世博上海案例馆 Expo Shanghai Pavilion
42	最佳BIM应用企业奖 Best enterprise for BIM application		上海现代建筑设计（集团）有限公司 Shanghai Xian Dai Architectural Design (Group) Co., Ltd.	
43			CCDI中建国际设计顾问有限公司 CCDI Group	
44			机械工业第六设计研究院 SIPPR Engineering Group Co., Ltd.	
45	BIM应用企业鼓励奖 Better enterprise for BIM application		天津水泥工业设计院有限公司 Tianjin Cement Industry Design and Research Institute Co., Ltd.	
46			沈阳市建筑设计院 Shenyang Architectural Design Institute	
47			华通设计顾问工程有限公司 Walton Design Consultant Engineering Co., Ltd.	
48			福建省建筑设计研究院 Fujian Architectural Design & Research Institute	
49			宝钢工程技术集团有限公司 Baosteel Engineering & Technology Group Co., Ltd.	
50			云南省设计院 Yunnan Design Institute	

2. 中国勘察设计协会《第二届"创新杯"BIM设计大赛（2011）》获奖名单
China Exploration & Design Association 〈 The 2nd "Innovation Cup" for BIM Design Competition (2011) 〉 Winner List

《第二届"创新杯"BIM设计大赛（2011）》竞赛情况 Summary of The 2nd "Innovation Cup" for BIM Design Competition (2011)			
竞赛时间 Competition Time	2011年8月 August, 2010	颁奖地点 Award prizes location	北京 Beijing
参赛单位数量 Number of Participants	67个 67	参赛项目数量 Number of projects	170个 170
获奖单位数量 Number of winners	21个 21	奖项数量 Amount of awards	44个 44

中国勘察设计协会《第二届"创新杯"BIM设计大赛（2011）》获奖名单
China Exploration & Design Association 〈 The 2nd " Innovation Cup" for BIM Design Competition (2011) 〉 Winner List

	项目分类 Project classification		奖项 Award	单位名称 Organization name	项目名称 Project name
1	最佳BIM建筑奖 Best BIM for architectural design		一等奖 First Prize	中国建筑设计研究院 China Architecture Design & Research Group	奥林匹克公园瞭望塔 Olympic Park Observation Tower
2			二等奖 Second Prize	现代设计集团上海建筑设计研究院有限公司 Xiandai Group Shanghai Architectural Design & Research Institute Co., Ltd	黑瞎子岛植物园 Heilongjiang Heixiazi Island Botanical Garden
3				中国建筑设计研究院 China Architecture Design & Research Group	龙岩金融商务中心（B地块）B3#楼 Longyan Financial and Business Centre

（续表1）

项目分类 Project classification		奖项 Award	单位名称 Organization name	项目名称 Project name
4	最佳BIM建筑奖 Best BIM for architectural design	三等奖 Third Prize	现代设计集团上海建筑设计研究院有限公司 Xiandai Group Shanghai Architectural Design & Research Institute Co., Ltd.	黑龙江五大连池火山博物馆 Heilongjiang Five Lake Volcano Museum
5			CCDI中建国际设计顾问有限公司 CCDI Group	中钢国际广场 Sinosteel International Plaza
6			云南省设计院 Yunnan Design Institute	园博会建筑项目 Yunnan Horticultural Exposition Project
7	最佳BIM工程设计奖 Best BIM for engineering design	一等奖 First Prize	中国建筑设计研究院 China Architecture Design & Research Group	龙岩金融商务中心（B地块）B3#楼 Longyan Financial and Business Centre
8		二等奖 Second Prize	现代设计集团上海建筑设计研究院有限公司 Xiandai Group Shanghai Architectural Design & Research Institute Co., Ltd	黑瞎子岛植物园 Heixiazi Island Botanical Garden in Heilongjiang Province
9			CCDI中建国际设计顾问有限公司 CCDI Group	哈尔滨西火车站东广场地下枢纽项目 Harbin West Railway Station underground hub

（续表2）

项目分类 Project classification		奖项 Award	单位名称 Organization name	项目名称 Project name
10	最佳BIM工程设计奖 Best BIM for engineering design	三等奖 Third Prize	上海上安机电设计事务所有限公司 Shanghai Shangan M&E Design Office Co., Ltd.	上海中山医院肝肿瘤及心血管综合楼 Shanghai Zhongshan Liver Cancer and Cardiovascular Hospital Building
11			现代设计集团华东建筑设计研究院有限公司 Xiandai Group East China Architectural Design & Research Institute Co., Ltd.	申都大厦改建工程 Shanghai Shendu Renovation project
12			上海现代工程咨询有限公司 Shanghai Xiandai Engineering consultant co., Ltd.	曹妃甸国际生态城水上会馆 Caofeidian International Eco-City Water Hall
13	最佳BIM协同设计奖 Best BIM for collaboration design	一等奖 First Prize	现代设计集团上海建筑设计研究院有限公司 Xiandai Group Shanghai Architectural Design & Research Institute Co., Ltd.	黑瞎子岛植物园 Heilongjiang Heixiazi Island Botanical Garden
14		二等奖 Second Prize	上海现代工程咨询有限公司 Shanghai Xiandai Engineering consultant co., Ltd.	曹妃甸国际生态城水上会馆 Caofeidian International Eco-City Water Hall
15			广东省建筑设计研究院 Guangdong Architectural Design and Research Institute	珠江新城F2-4地块项目 Pearl River New City F2-4 Block Project

(续表3)

项目分类 Project classification		奖项 Award	单位名称 Organization name	项目名称 Project name
16	最佳BIM协同设计奖 Best BIM for collaboration design	三等奖 Third Prize	中国建筑设计研究院 China Architecture Design & Research Group	龙岩金融商务中心（B地块）B3#楼 Longyan Financial and Business Centre
17			福建省建筑设计研究院 Fujian Architectural Design & Research Institute	福州CBD万达商业广场 Fuzhou CBD Wanda Commercial Plaza
18			中国航空规划建设发展有限公司 China Aviation Planning and Construction Development Co., Ltd.	建川博物馆 航空三线建设馆 Jianchuan Museum
19	最佳BIM绿色分析应用奖 Best BIM for sustainable design	一等奖 First Prize	现代设计集团上海建筑设计研究院有限公司 Xiandai Group Shanghai Architectural Design & Research Institute Co., Ltd.	黑龙江五大连池火山博物馆 Heilongjiang Five Lake Volcano Museum
20		二等奖 Second Prize	华通设计顾问工程有限公司 Walton Design Consultant Engineering Co., Ltd.	中日·唐山曹妃甸生态工业园 Sino-Japanese.Tangshan Caofeidian Eco-Industrial Park
21			现代设计集团上海建筑设计研究院有限公司 Xiandai Group Shanghai Architectural Design & Research Institute Co., Ltd.	黑瞎子岛植物园 Heilongjiang Heixiazi Island Botanical Garden

(续表4)

项目分类 Project classification		奖项 Award	单位名称 Organization name	项目名称 Project name
22	最佳BIM绿色分析应用奖 Best BIM for sustainable design	三等奖 Third Prize	中国联合工程公司第三建筑工程设计研究院 China United Engineering Corporation	钱江世纪城龙达大厦 Qianjiang Century Ronda Building
23			中国建筑设计研究院 China Architecture Design & Research Group	雅世合金公寓1号楼项目 YashiHejin Apartment Building No.1
24			深圳市筑博工程设计有限公司 Zhubo Design Group Co., Ltd.	南宁规划展览馆 Nanning Planning Exhibition Hall
25	基础设施类 Best BIM for infrastructure	一等奖 First Prize	中国水电顾问集团昆明勘测设计研究院 Hydro China Kunming Engineering Corporation	阿海水电站 In Hydropower Station Design
26		二等奖 Second Prize	上海现代工程咨询有限公司 Shanghai Xiandai Engineering Consultant Co., Ltd.	上海新虹桥国际医学中心区域规划 Shanghai New Hongqiao International Medical Center Regional Planning
27		三等奖 Third Prize	中国水电顾问集团北京勘测设计研究院 Hydro China Beijing Engineering Corporation	山东文登抽水蓄能电站三维地质模型 3D Geologic Modeling in Wendeng Hydropower Station
28	工业工程类 Best plant design	一等奖 First Prize	天津水泥工业设计院有限公司 Tianjin Cement Industry Design and Research Institute Co., Ltd.	曲阳金隅水泥厂三维模型设计 QuyangJinyu Cement Plant

（续表5）

项目分类 Project clas-sification	奖项 Award	单位名称 Organization name	项目名称 Project name
29	二等奖 Second Prize	机械工业第六设计研究院 SIPPR Engineering Group Co., Ltd	陕西秦川机械发展股份公司恒温车间改造项目 Shaanxi Qinchuan Machinery Company Thermostat Workshop Renovation Project
30		上海上安机电设计事务所有限公司 Shanghai Shangan M&E Design Office Co., Ltd	虹桥商务核心区（一期）区域供能能源中心及配套工程 Hongqiao Business Core Area Power Center
31 工业工程类 Best plant design		机械工业第六设计研究院 SIPPR Engineering Group Co., Ltd	郑州煤矿机械公司高端液压支架生产基地建设项目 Zhengzhou Coal Mining Machinery Company production base of high-end hydraulic support
32	三等奖 Third Prize	上海现代建筑设计（集团）有限公司 Shanghai Xian Dai Architectural Design（Group）Co., Ltd.	世博变电站 Expo Transformer Substation
33		上海市政工程设计研究总院（集团）有限公司 Shanghai Municipal Engineering Design Institute Group	白龙港污泥处理工程污泥干化车间三维设计 Bailonggang Sludge Treatment Engineering Sludge Drying Workshop

(续表6)

项目分类 Project classification		奖项 Award	单位名称 Organization name	项目名称 Project name
34	最佳BIM应用企业奖 Best enterprise for BIM application		上海现代建筑设计（集团）有限公司 Shanghai Xian Dai Architectural Design（Group）Co., Ltd.	
35			中国建筑设计研究院 China Architecture Design & Research Group	
36			机械工业第六设计研究院 SIPPR Engineering Group Co., Ltd.	
37			中国水电顾问集团昆明勘测设计研究院 Hydro China Kunming Engineering Corporation	
38			CCDI中建国际设计顾问有限公司 CCDI Group	
39	BIM应用企业鼓励奖 Better enterprise for BIM application		天津水泥工业设计院有限公司 Tianjin Cement Industry Design and Research Institute Co., Ltd.	
40			上海上安机电设计事务所有限公司 Shanghai Shangan M&E Design Office Co., Ltd.	
41			云南省设计院 Yunnan Design Institute	
42			福建省建筑设计研究院 Fujian Architectural Design & Research Institute	
43			深圳市建筑设计研究总院有限公司 Shenzhen General Institute of Architectural Design and Research Co., Ltd.	
44	BIM应用特等奖 Special l award of BIM application	上海中心大厦 Shanghai Tower	上海中心大厦建设发展有限公司 Shanghai Tower Construction and Development Co., Ltd.	
			同济大学建筑设计研究院（集团）有限公司 Tongji Architectural Design (Group) Co., Ltd.	
			上海上安机电设计事务所有限公司 Shanghai Shangan M&E Design office Co., Ltd.	
			上海建工集团股份有限公司 Shanghai Construction (Group) Co., Ltd.	
			沈阳远大铝业工程有限公司 Yuanda China	

3. 中国勘察设计协会《第三届"创新杯"BIM设计大赛(2012)》获奖名单
China Exploration & Design Association 〈 The 3nd "Innovation Cup" for BIM Design Competition (2012) 〉 Winner List

《第三届"创新杯"BIM设计大赛(2012)》竞赛情况 Summary of The 3nd "Innovation Cup" BIM Design Competition (2012)			
竞赛时间 Competition Time	2012年8月 August, 2010	颁奖地点 Award prizes location	北京 Beijing
参赛单位数量 Number of organization Participants	76个 76	参赛项目数量 Number of projects	198个 198
获奖单位数量 Number of winners	27个 27	奖项数量 Amount of awards	48个 48

中国勘察设计协会《第三届"创新杯"BIM设计大赛(2012)》获奖名单
China Exploration & Design Association 〈 The 3nd "Innovation Cup" for BIM Design Competition (2012) 〉 Winner List

	项目分类 Project classification	奖项 Award	单位名称 Organization name	项目名称 Project name
1	最佳BIM建筑设计奖 Best BIM for Architectural Design	一等奖 First Prize	北京市建筑设计研究院 Beijing Institute of Architectural Design	珠海歌剧院 Zhuhai Opera House
2		二等奖 Second Prize	现代设计集团上海建筑设计研究院有限公司 Xiandai Group Shanghai Architectural Design & Research Institute Co., Ltd	沈阳文化艺术中心 The Shenyang Cultural Arts Center
3			北京市建筑设计研究院 Beijing Institute of Architectural Design	重庆国际马戏城 Chongqing International Circus

(续表1)

项目分类 Project classification		奖项 Award	单位名称 Organization name	项目名称 Project name
4	最佳BIM建筑设计奖 Best BIM for Architectural Design	三等奖 Third Prize	香港华艺设计顾问（深圳）有限公司 Huayi Design Consultant Group	济南中海广场—寰宇城 Jinan Zhonghai Square - Universal City
5			现代设计集团华东建筑设计研究院有限公司 Xiandai Group East China Architectural Design & Research Institute Co. Ltd.	南京禄口国际机场二期工程 Nanjing Lukou International Airport Second Phase
6			深圳市建筑设计研究总院有限公司 Shenzhen General Institute of Architectural Design and Research Co., Ltd.	能源大厦 Energy Mansion
7	最佳BIM工程设计奖 Best BIM for Engineering	一等奖 First Prize	上海现代建筑设计集团工程建设咨询有限公司 Shanghai Xian Dai Architecture, Engineering & Consulting Co., Ltd.	思南路旧房改造—多维技术在历史保护建筑群中的作用 Si'nan Ancient Buildings Renovation in Shanghai
8		二等奖 Second Prize	上海建筑设计研究院有限公司 Shanghai Architectural Design & Research Institute Co., Ltd.	长征医院浦东新院 Changzheng Hospital Pudong New Building
9			北京市建筑设计研究院有限公司 Beijing Institute of Architectural Design	绍兴体育中心 Shaoxing Sports Center

(续表2)

项目分类 Project classification	奖项 Award	单位名称 Organization name	项目名称 Project name
10	最佳BIM工程设计奖 Best BIM for Engineering	三等奖 Third Prize	现代设计集团华东建筑设计研究院有限公司 Xiandai Group East China Architectural Design & Research Institute Co., Ltd.	南京禄口国际机场二期工程 Nanjing Lukou International Airport Second Phase
11			中国航空规划建设发展有限公司 China Aviation Planning and Construction Development Co., Ltd.	宝苑住宅小区二期工程 Baoyuan Residential District Second Phase
12			中建国际（深圳）设计顾问有限公司 CCDI Group 中国建筑第八工程局有限公司 China Construction Eighth Engineering Bureau Co., Ltd.	毛里求斯机场建设项目 Mauritius Airport
13	最佳BIM协同设计奖 Best BIM for Collaboration	一等奖 First Prize	中国建筑设计研究院建筑专业设计研究院 CADRG China Architectural Design & Research Group	中国移动国际信息港二期A标段（研发创新中心） China Mobile International Information Port Second Phase (R & D Innovation Center)
14		二等奖 Second Prize	中国航空规划建设发展有限公司 China Aviation Planning and Construction Development Co., Ltd.	中国航空规划公司科研综合楼 Chinese Aviation Planning Company Research Building
15			上海市城建（集团）公司 Shanghai Urban Construction Group 上海市地下空间设计研究总院有限公司 Shanghai Underground Space Architectural Design & Research Institute Co., Ltd.	BIM在上海城建浦江基地预制装配式住宅工程中的应用 Pujiang Base Residential Project in Shanghai

（续表3）

	项目分类 Project classification	奖项 Award	单位名称 Organization name	项目名称 Project name
16	最佳BIM协同设计奖 Best BIM for Collaboration	三等奖 Third Prize	中建国际（深圳）设计顾问有限公司 CCDI Group 中国建筑第八工程局有限公司 China Construction Eighth Engineering Bureau Co., Ltd.	毛里求斯机场建设项目 Mauritius Airport
17			山东同圆设计集团有限公司 Shandong Tongyuan Design Group	省会文化艺术中心"三馆"项目 Jinan Cultural Arts Center Project (Three Venues)
18			上海建筑设计研究院有限公司 Shanghai Architectural Design & Research Institute Co., Ltd.	沈阳文化艺术中心 Shenyang Cultural Arts Center
19	最佳BIM绿色分析应用奖 Best BIM for Sustainable Design	一等奖 First Prize	天津市建筑设计院 Tianjin Architectural Design Institute	解放南路文体中心 Jiefang South Road Culture and Sports Center
20		二等奖 Second Prize	现代设计集团华东建筑设计研究院有限公司 Xiandai Group East China Architectural Design & Research Institute Co. Ltd.	南京禄口国际机场二期 Nanjing Lukou International Airport Second Phase
21			中国联合工程公司 China United Engineering Corporation	"杭州之门"项目绿色设计分析 Hangzhou's Gate

(续表4)

项目分类 Project classification	奖项 Award	单位名称 Organization name	项目名称 Project name
22	最佳BIM绿色分析应用奖 Best BIM for Sustainable Design	三等奖 Third Prize	深圳市建筑设计研究总院有限公司 Shenzhen General Institute of Architectural Design and research Co., Ltd.	能源大厦 Energy Building
23			中国建筑设计研究院 China Architectural Design and Research Group	中国建筑设计研究院创新科研示范楼 CAG Innovative Research and Demonstration Building
24			中国电子工程设计院 China Electronics Engineering Design Institute	中国华融大厦 China Huarong Building
25	基础设计类 Best BIM for Infrastructure	一等奖 First Prize	中国水电顾问集团昆明勘测设计研究院 Hydro China Kunming Engineering Corporation	黄登水电站施工总布置BIM协同设计 Huangdeng Hydropower Station
26			中国中铁二院工程集团有限责任公司 China Railway Eryuan Engineering Group Co., Ltd.	西部某高速铁路三维设计 BIM Design for High-speed Railway in West China
27		二等奖 Second Prize	上海市政工程设计研究总院（集团）有限公司 Shanghai Municipal Engineering Design Institute (Group) CO., Ltd.	BIM在地下空间的应用 BIM Application in Underground Space Project

(续表5)

项目分类 Project classification	奖项 Award	单位名称 Organization name	项目名称 Project name
28	基础设计类 Best BIM for Infrastructure	三等奖 Third Prize	上海建筑设计研究院有限公司 Shanghai Institute of Architectural Design & Research	大连专用车产业科技创新基地 Dalian Special Vehicle Industry Technology Innovation Base
29			上海市地下建筑设计研究总院 Shanghai Underground Space Architectural Design and Research Institute Co., Ltd.	使用BIM技术在地铁车站中的实践 Practice of using BIM Technology in Subway Station
30	工业工程类 Best Plant Design	一等奖 First Prize	机械工业第六设计研究院有限公司 SIPPR Engineering Group Co., Ltd.	河南中烟公司许昌卷烟厂易地技术改造项目 Henan Tobacco Company, Xuchang Cigarette Factory Technological Transformation Projects
31		二等奖 Second Prize	中国电力工程顾问集团西南电力设计院有限公司 CPECC Southwest Electric Power Design Institute Southwest Electric Power Design Institute	沙州750kV变电站 Shazhou 750KV Transformer Substation
32		三等奖 Third Prize	东风设计研究院有限公司 Dongfeng Design Institute Group	CAP污水处理站总承包项目 CAP Sewage Treatment Station EPC Project

（续表6）

项目分类 Project classification	奖项 Award		单位名称 Organization name	项目名称 Project name
33	最佳BIM拓展应用奖 Best BIM Integration on Application	民用建筑类 Building	中建三局第一建设工程有限责任公司 The First Construction Engineering Limited Company of China Construction Third Engineering Bureau	嘉里建设广场二期 Kerry Plaza Second Phase
34			中国建筑设计研究院建筑专业设计研究院 China Architecture Design & Research Group	中国移动国际信息港二期A标段（研发创新中心） China Mobile International Information Port Phase II (R & D Innovation Center)
35			上海城建（集团）公司 Shanghai Urban Construction (Group) Corporation 上海市地下空间设计研究总院有限公司 Shanghai Underground Space Architectural Design & Research Institute Co., Ltd.	BIM在上海城建浦江基地预制装配式住宅工程中的应用 Pujiang Base Residential Project in Shanghai
36			上海现代建筑设计集团工程建设咨询有限公司 Shanghai Xian Dai Architecture, Engineering & Consulting Co., Ltd.	思南路旧房改造—多维技术在历史保护建筑群中的作用 Si'nan Road Ancient Buildings Renovation in Shanghai
37		基础设施类 Infrastructure	铁道第三勘察设计院集团有限公司 The Third Railway Survey and Design Institute Group Corporation	心站—高铁客站BIM设施运营管理系统 Railway Passenger Station Facility Management System

(续表7)

	项目分类 Project classification		奖项 Award	单位名称 Organization name	项目名称 Project name
38	基础设施类 Infrastructure		工业工程类 Plant	机械工业第六设计研究院有限公司 SIPPR Engineering Group Co., Ltd.	郑州热力总公司南郊热源厂集中供热工程 Zhengzhou Heating Company Central Heating Plant Project
39		最佳BIM应用企业奖 Best Enterprise for BIM Application		上海现代建筑设计（集团）有限公司 Shanghai Xian Dai Architectural Design （Group）Co., Ltd.	
40				北京市建筑设计研究院 Beijing Institute of Architectural Design	
41				中国建筑设计研究院 China Architecture Design & Research Group	
42				中国航空规划建设发展有限公司 China Aviation Planning and Construction Development Co., Ltd.	
43				中国水电顾问集团昆明勘测设计研究院 Hydro China Kunming Engineering Corporation	
44				机械工业第六设计研究院有限公司 SIPPR Engineering Group Co., Ltd.	
45				深圳市建筑设计研究总院有限公司 Shenzhen General Institute of Architectural Design and research Co., Ltd.	
46				香港华艺设计顾问（深圳）有限公司 Huayi Design Consultant Co., Ltd.	
47				福建省建筑设计研究院 Fujian Architectural Design & Research Institute	
48				中建国际（深圳）设计顾问有限公司 CCDI Group	

注：

1. 获奖单位和获奖项目中文名称来自中国勘察设计协会获奖名单。

2. 获奖单位和获奖项目英文名称来自各单位官方网站。

Annotation：

1. The Chinese names of the winners and projects are from China Exploration & Design Association.

2. The English names of the winners and projects are from official websites of each organization.

1.3 BIM技术的主要特点
Major Features of BIM Technology

1. 特点之一：可视化设计
Feature 1: Visualization design

基于BIM设计成果的效果图、虚拟漫游、仿真模拟等多种项目展示手段，可以让各参与方对项目本身进行深度直观的了解。

BIM-based project presentation methods such as architectural renderings, virtual walkthroughs and simulation can help project participants have profound and visual understanding of the project.

2. 特点之二：参数化设计
Feature 2: Parametric design

参数化设计的意义在于将建筑构件和设备的各种真实属性通过参数的形式进行模拟，并进行相关的数据统计和模拟分析计算。通过参数调整，可驱动构件形体发生改变以及性能模拟比较，满足设计要求。

The vital aspect of parametric design is to simulate various actual attributes of building components and facilities through parameters, and be able to make related data statistics, simulation analysis and calculations. Through parameter adjustments, it can manipulate the component shape to change and compare performance and satisfy design requirements.

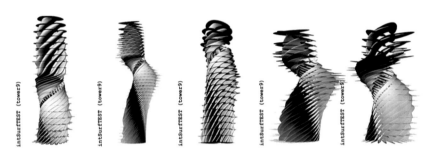

3. 特点之三：关联修改设计
Feature 3: Correlation modification

BIM模型的几何与参数联动特性，使得模型与模型之间、模型与视图之间、模型与统计数据之间保持实时关联，从而实现一处修改处处更新，提高设计和修改效率。

The geometric and parametric association of BIM models can realize real time correlations between models, model and view, and model and statistical data, which make it possible to upgrade everything by changing one aspect, thus improving design and modification efficiency.

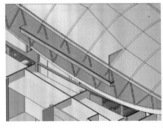
站位控制面（黄色）

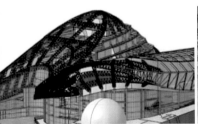
屋面钢结构

钢结构局部1

4. 特点之四：任务划分与管理
Feature 4: Task partitioning and management

设计工作重心前置，重新优化各专业间的工作界面，同时优化管理效率和管理流程，增强项目风险控制能力，实现精细化管理。

BIM can realize focus preposition of design, re-optimize the work interface between different specialties, optimize management efficiency and workflow, consolidates project risk control and realizes refined management.

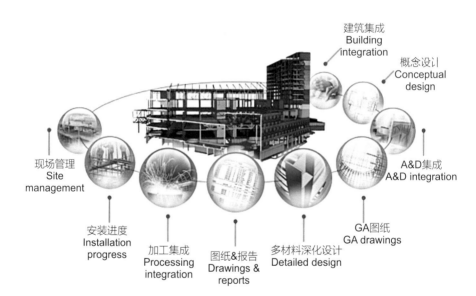

5. 特点之五：性能分析
Feature 5: Performance analysis

基于BIM设计成果的光照、能耗、风环境、消防疏散、可视度等建筑性能分析，为设计优化提供了依据，消除未来使用中可能存在的隐患。

BIM-based building performance analysis of lighting, energy consumption, wind environment, fire escape, visibility, etc., can support design optimization and eliminate any possible future risks.

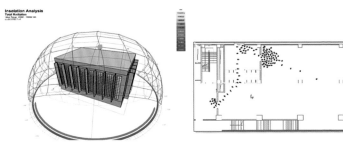

6. 特点之六：协同设计
Feature 6: Collaborative design

基于相同BIM设计平台的多专业多团队协同设计工作模式，辅以实时协同、阶段性协同、三维校审的工作方法，及时解决各种错漏碰缺，提高设计质量。

Multidisciplinary and multi-group collaborative design based on BIM, supported by real time collaboration, phased collaboration and 3D review, can solve various errors and flaws and improve design quality in a timely manner.

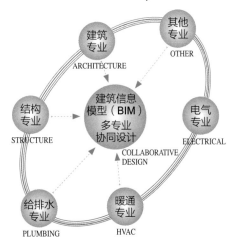

7. 特点之七：三维设计交付
Feature 7: 3D design delivery

目前基于BIM模型生成的高质量二维施工图纸，以及全套BIM设计及浏览模型的三维设计交付模式，将成为未来全三维BIM交付模式的过渡阶段。

The current high quality BIM-based 2D construction drawings, as well as 3D design delivery mode including BIM design and browse models, will become a transition stage for all 3D BIM delivery modes in the future.

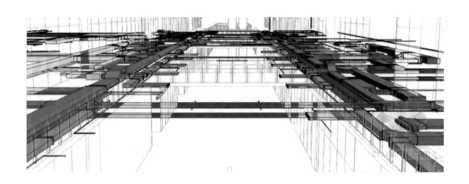

8. 特点之八：远程与移动平台工作
Feature 8: Remote and mobile platform work

基于互联网及云技术的移动终端（智能手机、平板电脑等）和数据管理平台，打破了空间、地域对设计工作的限制，提高了工作效率。

Internet and cloud technology-based mobile terminals (such as smart phones and tablets) and data management platforms break through spatial and territorial restrictions to enhance design and work efficiency.

1.4 BIM技术未来的发展趋势
Future Trends of BIM Technology

1. **趋势之一：国家发展目标与BIM未来技术发展相一致**
   Trend 1: Consistency of national development goals and future BIM technological development

2011年7月6日住房与城乡建设部颁布《建筑业发展十二五规划》
The 12th Five-Year Plan for Construction Industry issued by Ministry of Housing and Urban-Rural Development on July 6, 2011

"十二五"期末，努力实现"技术进步"目标
By the end of the 12th Five-Year Plan, to realize the goal of "technical improvement"

- 大型骨干工程设计企业基本建立协同设计、三维设计的设计集成系统。
- 大型骨干勘察企业建立三维地层信息系统。

- Large key design enterprises should establish a design integration system including collaborative design and 3D design.
- Large key engineering survey enterprises should build a 3D stratum information system.

加强技术进步和创新的主要措施
Main measures of consolidating technical improvement and innovation

- 全面提高行业工业化、信息化、城镇化水平。
- 建立涵盖设计、施工全过程的信息化标准体系。
- 加快关键信息化标准的编制。
- 促进行业信息共享。

- Improve the overall level of industrialization, information and urbanization.
- Build a standard system of information technology covering the whole process of design and construction.
- Speed up preparation of key information technology standards.
- Promote sharing of industrial information.

2. 趋势之二：未来BIM的发展整体架构图
Trend 2: Overall framework of future BIM development

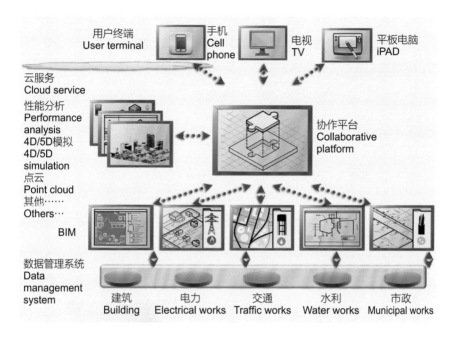

3. 趋势之三：BIM技术促进了决策流程和成本控制的优化
Trend 3: BIM technology optimizes decision-making flow and cost control

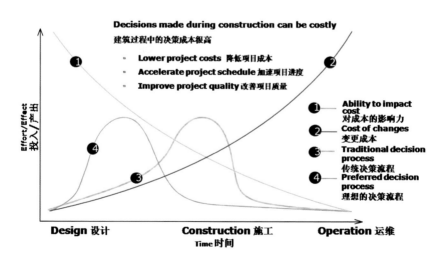

4. 趋势之四：BIM技术应用的高价值体现
Trend 4: High value of BIM application

BIM应用的两个维度
Two Dimensions of BIM Application

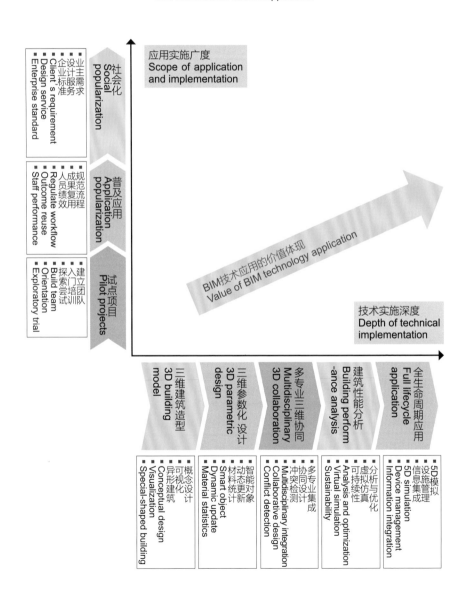

5. 趋势之五： 5D技术对项目成本、周期、质量的影响力
Trend 5: Influence of 5D technology on project cost, cycle and quality

3D设计模型基础上增加施工进度（4D -Time）及成本（5D -Cost）
3D design model is added with 4D-Time and 5D-Cost.

5D技术的四大目标
Four Goals of 5D Technology

- 节省5%-15%的建造成本
  To save 5%-15% construction cost
- 缩短5%-15%的项目时间
  To reduce 5%-15% project duration
- 提高20%-30%的项目质量
  To improve 20%-30% project quality
- 降低决策风险，提高投资效益
  To minimize decision-making risk and increase investment benefit

6. 趋势之六：云计算对建筑产业发展的影响
Trend 6: Influence of cloud computing on the development of the construction industry

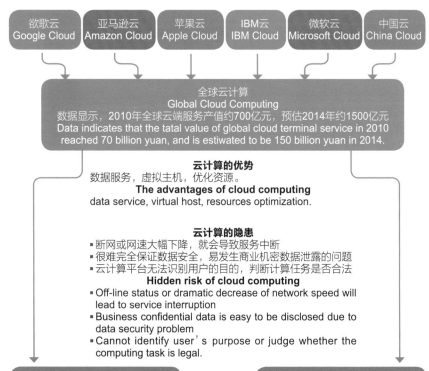

信息产业三大核心发展方向
1. 网络运输能力
2. 中央"云"处理能力
3. 新终端的设计能力

Three Major Developing Trends of the Information Industry
1. Network transport capacity
2. Central "cloud" processing capacity
3. Design capacity of new terminals

7. 趋势之七：BIM技术对智能建筑及数字城市的技术支撑
Trend 7: Technical support of BIM technology in smart buildings and digital cities

基于BIM技术、GIS系统构建智能化（物联网）建筑及数字化城市管理系统。

Build smart (the Internet of Things) building and digital city management system based on BIM technology and GIS system.

8. 趋势之八：绿色可持续及装配式建筑设计
Trend 8: Green sustainability and fabricated architectural design

基于BIM技术的绿色节能分析，绿色建筑认证，基于BIM的构件装配式生产体系，降低成本，保证质量。

BIM-based green energy-saving analysis, green building certification, and BIM-based fabricated production system can reduce cost and assure quality.

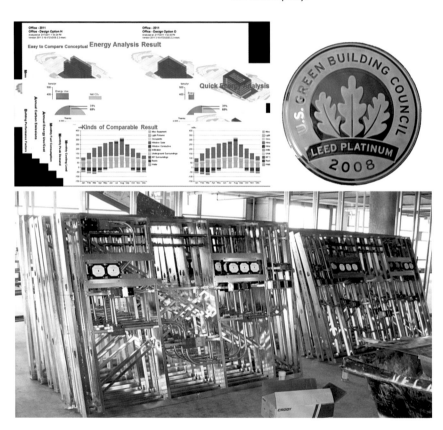

参考文献

1. 住房和城乡建设部《2011—2015年建筑业信息化发展纲要》
2. McGraw Hill公司发布《北美BIM商业价值评估报告（2007—2012）》

References

1. Guidelines for Information Development of the Construction Industry 2011－2015 published by Ministry of Housing and Urban-Rural Development
2. The Business Value of BIM in North America (2007－2012) published by McGraw-Hill Construction.

BIM技术应用的优势和劣势比较

Comparison of Application of the Advantages and Disadvantages of BIM Technology

2.1 二维设计软件与BIM设计软件的十四个比较
14 Top Differences Between 2D Design Software and BIM Design Utilities

设计信息在整个设计过程中的传递关系
Delivery of design information between project stages and disciplines

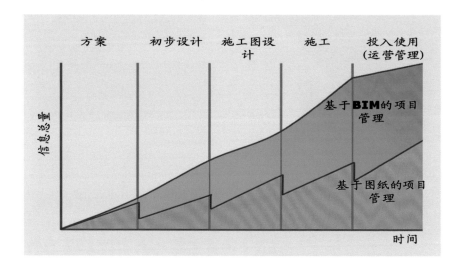

传统二维设计软件
2D design software

> 以图纸为中心,数据在工程不同阶段及不同专业间的传递会有损失或失真。
> Because the focus in 2D design is on drawing, information is easily lost and distorted when delivered between different stages and different disciplines.

BIM设计软件
BIM design software

> 以工程信息为中心,BIM信息在工程任何阶段和各专业间传递是连续和无损失的。
> Because of project focus, BIM information can be delivered without any loss or distortion, between different stages and different disciplines.

对比结果
Conclusion

二维设计软件的信息在不同工程阶段及不同专业间的传递有损失;而三维BIM设计软件可实现信息更有效地传递!

Information is easily lost when delivering information in 2D design software, between different stages and different disciplines, but 3D BIM software can deliver the information more efficiently!

设计工作量、设计过程的重心变化关系
Change of focus in terms of design workload and design process

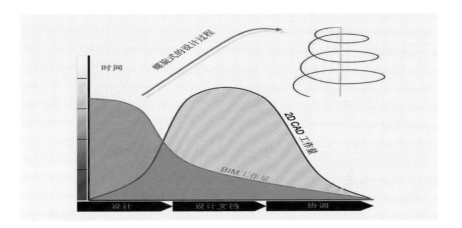

传统二维设计软件
2D design software

- 设计人员花费大量时间在制图和协调上，在关系到设计质量的方案设计、专业技术设计等核心工作的投入不足，造成后期修改工作量增加。
- Coordination requires a great deal of time and resources, often at the cost of important factors affecting design quality, such as MEP design and calculation. This will increase workload in later project stages.

BIM设计软件
BIM design software

- 从模型自动生成的图纸，提高设计效率、质量，使设计师更关注方案比选、专业技术设计、优化、协同等核心工作，减少后期的重复设计工作量。
- Plan drawings can be generated automatically from the BIM model. This improves design efficiency and design quality, enabling the designers to focus on core tasks, such as design comparison, calculation, optimization and coordination. This reduces workload in later project stages.

对比结果
Conclusion

为保证设计质量，二维设计软件设计人员的大量时间花在协调和对图上，后期修改工作量大；而三维BIM设计软件设计工作量前移，重点在方案比选和技术优化，一旦模型关联关系建立，修改便利！

To assure the design quality, the 2D designers spend much time in coordinating and comparing drawing, and will suffer heavy workload if any change in drawing after; however, the 3D BIM design software can put the work in advance, and focus on the scheme choice and technology optimization. When the model is setup, it is easy to make change.

计算与绘图的融合修改关系
Integration of calculation and drafting

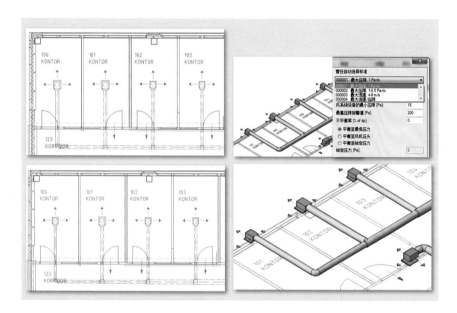

传统二维设计软件
2D design software

> 计算与绘图脱节，图形与计算结果不能双向更新。
> Because calculation and drafting are separated, drawings and calculation results cannot be synchronized.

BIM设计软件
BIM design software

> 计算与绘图融合，图形与计算保持实时关联、自动更新，同时支持人工干预和调整。
> Calculation and drafting can be combined. Drawings and calculation results can be updated in real time. Customization can also be supported.

对比结果 Conclusion

二维设计软件的专业计算基本与绘图脱节；而三维BIM设计软件将计算与绘图融合，做到一处修改，处处更新。

Calculation and drafting are separated by 2D design software. They can be combined by BIM design software, enabling updates in real time.

BIM技术应用的优势和劣势比较
Comparison of Application of the Advantages and Disadvantages of BIM Technology

二维图块与真实产品数据库的关系
Relationship between 2D symbols and real products

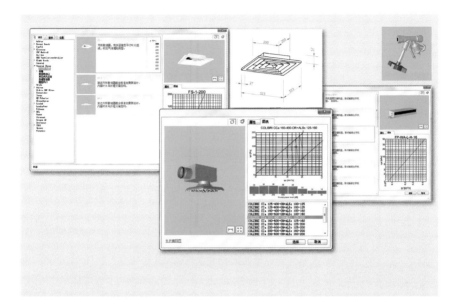

传统二维设计软件
2D design software

> 机电专业的CAD的图块仅表达外形，设备数据信息少
> AutoCAD 2D symbol blocks in MEP can only include geometry without much technical data.

BIM设计软件
BIM design software

> BIM机电构件设备外形与真实产品的完整数据信息，并支持用户编辑和扩充，并可运行工况的模拟。
> BIM MEP include geometry and complete technical data which enable project progress simulations and customization.

C对比结果
onclusion

二维设计软件CAD图块仅表达外形，相关数据缺失；而三维BIM设计软件可提供"数形合一"的构件库！

AutoCAD 2D blocks created by 2D design software include only geometry without much technical data. With BIM design software, product databases with real products can contain both geometry and technical data.

53

设备数据与工作状态模拟的关系
Relationship between technical product details and project progress simulations

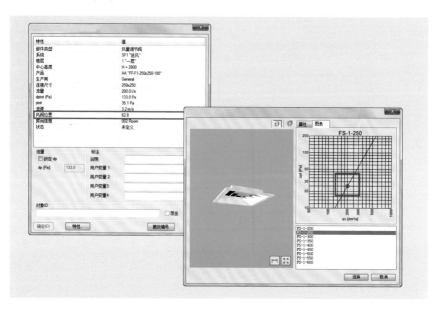

传统二维设计软件
2D design software

> 由于设备图块无数据信息,所以无法进行工作状态模拟。
> The project progress cannot be simulated using 2D design software because 2D blocks do not contain technical data.

BIM设计软件
BIM design software

> 由于设备构件带有完整的数据信息,因此可进行校核计算,必要时能重新进行管径选择和设备选型;同时可进行工作状态模拟。
> Databases with real products include complete technical data from manufacturers, which can be used for calculations such as system sizing and equipment selection, and for simulating project progress.

对比结果 Conclusion

二维设计软件无法进行设备工作状态模拟;而三维BIM设计软件通过设备构件信息,可准确模拟设备的工作状态!

Project progress simulations are not supported by 2D design software, but they can be supported by BIM design software. This is made possible by product databases that contain complete technical data on real, commercially available products.

平面、立面及剖面的对应关系
The corresponding relationship between plane, elevation and section

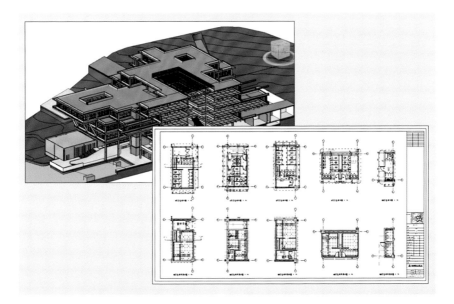

传统二维设计软件
2D design software

> 平面、立面及剖面之间互相割裂，导致视图的创建和修改工作量巨大，不能进行关联修改。

> As plane, elevation and section are separated, it is not possible to associate and edit them in real time. This will increase workload, e.g. creating viewports and editing.

BIM设计软件
BIM design software

> 由模型自动生成所有视图，并相互关联。可根据设计需要，任意创建多点剖面，提高设计质量和效率。

> Viewports can be created automatically and they can be associated with BIM model. Designers can create as many sections as they need in order to improve design efficiency and design quality.

对比结果
Conclusion

二维设计软件的平面、立面和剖面相互对应是一个难题；而BIM设计软件可以实现模型和视图之间自动关联与更新！

Associating plane, elevation and section is quite difficult using 2D design software. BIM design software makes it possible to associate and update them automatically.

机电管线碰撞检测与综合的关系
MEP pipeline collision detection and pipeline integration

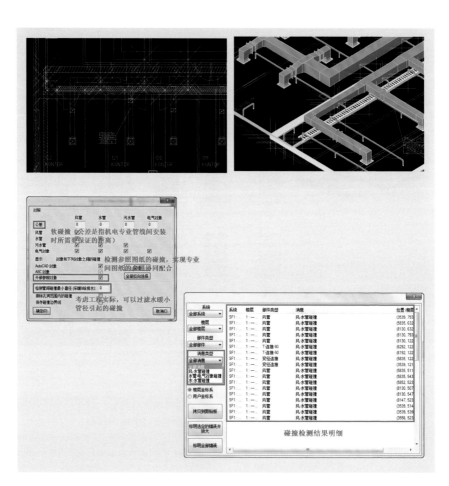

对比结果
Conclusion

二维设计软件的机电管线难以实现碰撞检测；而BIM设计软件通过机电管线综合与碰撞检测，实现高效、高质量的协同设计！

Efficient collision detection between MEP pipelines cannot be achieved using 2D design software. BIM design software offers MEP pipeline collision detection and pipeline integration, making real-time coordination easy, efficient and high quality.

机电管线与建筑结构的配合关系
Coordination between MEP pipelines and construction

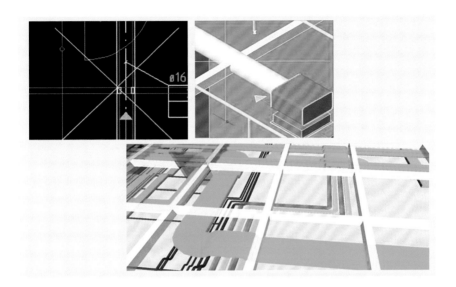

传统二维设计软件
2D design software

> 机电与建筑结构专业之间，通过人工方式进行专业配合，很难避免设计中的冲突和碰撞问题，一些问题在施工阶段才暴露。

> Coordination between MEP pipelines and construction can only be achieved manually. Because of this, many collisions are not noticed before the construction stage.

BIM设计软件
BIM design software

> BIM设计软件的自动碰撞检测、实时和阶段协同设计功能，将各种冲突和碰撞问题，消灭在设计阶段，并实现了空间的高效利用。

> Collisions can be resolved in the design stage with automatic collision detection and real-time coordination provided by the BIM design software. This also enables more efficient use of space.

对比结果
Conclusion

二维设计软件的很容易产生碰撞，并浪费建筑有效空间；而BIM设计软件通过专业协同设计，减少碰撞，实现建筑空间的有效利用！

With 2D design software, many collisions are not noticed before the construction stage and a great deal of space will be wasted. BIM design software enables coordination between different disciplines, making it possible to resolve collisions and use space more efficiently.

对比9 机电管线预留洞与土建预留洞的配合关系
Coordination between provisions for voids of MEP and construction

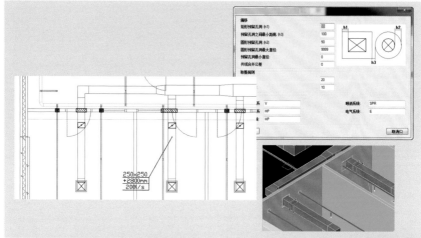

孔洞预留的展示

传统二维设计软件
2D design software

- 土建的机电预留洞数量繁多，尺寸和位置多样，易造成错误和遗漏。
- MEP pipelines require a large number of provisions for voids with different dimensions and locations, making errors very common when using 2D design software.

BIM设计软件
BIM design software

- BIM设计软件提供了预留洞的功能，土建专业可根据机电专业的碰撞和管线综合结果，实现预留洞的自动创建、更新和专业预留洞位置关系检查。
- BIM design software provides a Provision for Voids function that allows construction designers to create and update provisions for voids for MEP pipelines automatically, based on the results of MEP pipeline integration. It is also possible to check the location relationships between the provisions.

对比结果
Conclusion

二维设计软件的机电管道与土建预留洞经常冲突和不一致；而BIM设计软件通过专门的预留洞功能，实现实时的预留洞协同设计！

Using 2D design software causes many differences and collisions between MEP pipelines and provisions for voids in the construction stage. Real-time coordination can be achieved using the Provision for Voids function provided by BIM design software.

对比 10 设计信息与图形的融合关系
Integration of design information and model

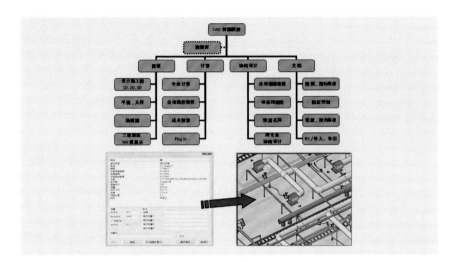

传统二维设计软件
2D design software

> 因为所有图形都是由CAD点、线、圆组成，通过几何信息和图层、扩展信息叠加而成，故信息容易丢失。

> All drawings are made by basic CAD objects, such as point, line and circle. Information is stacked by geometry, layer and extended information. Because of this, information is easily lost between project stages.

BIM设计软件
BIM design software

> 模型中的专业图元是构件实体在软件中的虚拟表现，含有大量的专业信息，故信息不易丢失。

> Objects in BIM model are exact replicas of real products from real manufacturers. Each object contains complete technical data and information, which cannot be lost between project stages.

对比结果
Conclusion

二维设计软件的CAD是"点、线、圆"组合和叠加，信息易丢失；而三维BIM设计软件是实体在软件中的虚拟模型，信息完整！

With 2D design software, information is easily lost between project stages because drawings are created using basic CAD objects without any technical data. This will not happen with BIM design software because BIM models contain complete technical data.

对比 11　一维、二维、三维的信息融合关系
Integration of information between 1D, 2D and 3D drawings

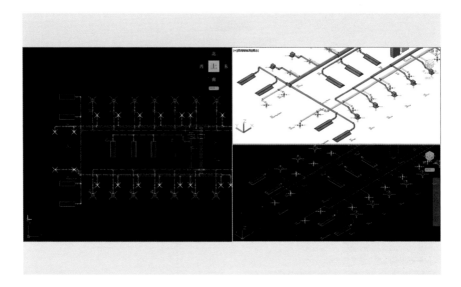

传统二维设计软件
2D design software

- 一维图（原理图、系统图）必须手工绘制，与已绘制的二维图（平、立、剖面）没任何联系。
- 1D drawings (such as isometric drawings and schematic drawings) can only be made manually. There is no relationship between 1D drawing and 2D drawings (such as plane, elevation and section drawings).

BIM设计软件
BIM design software

- BIM技术使一维、二维、三维（模型）及信息完全融合成为可能。
- Information can be combined between 1D, 2D, and 3D drawings using BIM design software.

对比结果
Conclusion

　　二维设计软件绘制一维图和二维图没有任何联系；而BIM设计软件使一维、二维、三维及信息完全融合成为可能！

　　In 2D design software, there is no relationship between 1D drawings and 2D drawings; BIM design software allows information to be combined between 1D, 2D, and 3D drawings.

对比 12 工程量及材料数量统计的准确性关系
Exact quantities and volumes of construction and bills of materials

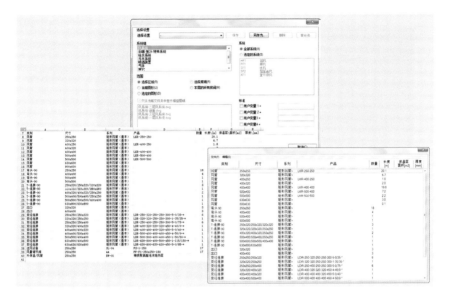

传统二维设计软件
2D design software

> 人工方式无法按系统、跨图纸进行工程量、材料数量统计，准确性偏低。
> Exact quantities and volumes in construction and bills of materials cannot be counted because of different systems and different drawings.

BIM设计软件
BIM design software

> 由于三维模型包含完整的数据信息，所以可按需求进行各种工程量、材料数量统计，自动生成材料表等，统计信息准确可信。
> Designers can get exact quantities and volumes when they want and bills of materials can be created automatically. Because the BIM model contains complete product information, the results are reliable.

对比结果 Conclusion

二维设计软件的工程量及材料量的统计准确度偏低；而BIM设计软件的统计信息是从虚拟模型中提取的，准确可信！

It is impossible to count exact quantities and volumes in construction and bills of materials using 2D design software. BIM design software can do this automatically and reliably.

对比 13 设计质量的对比关系
design quality

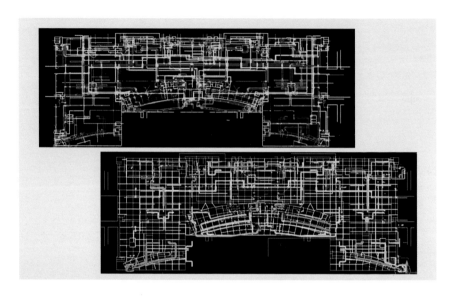

传统二维设计软件
2D design software

> 由于专业间配合不到位、信息与图形分离，易产生施工图隐患，降低设计质量。

> Hidden problems will be left in plan drawings, because coordination is not in place and information and drawings are separated. This will reduce the quality of design.

BIM设计软件
BIM design software

> BIM设计软件通过专业间的实时协同设计，解决了上述隐患，实现精细化设计，提高了设计质量。

> Hidden problems can be resolved by real-time coordination in BIM design software. This makes it possible to achieve precise design and better quality of design.

对比结果
Conclusion

二维设计软件受功能所限，很难在一般性层面设计质量上有所突破；而BIM设计软件从协同设计上保证了设计质量，使设计结果更加可靠！

The limitations of 2D design software make it impossible to improve design quality on general level. Quality of design can be assured by coordination in BIM design software and the results will be more reliable.

对比 14 设计信息的流动与传递关系
Flow and delivery of design information

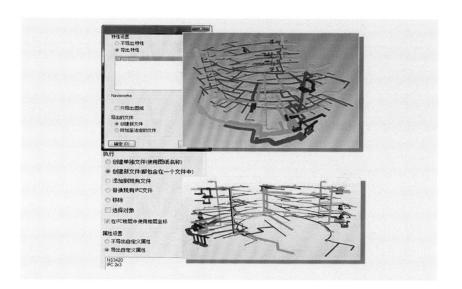

传统二维设计软件
2D design software

- 因各专业采用不同的技术软件，专业之间几乎没有任何设计信息的传递，只是最基础的二维图纸的配合。
- Design information cannot be delivered effectively between different disciplines. This is because different disciplines use different technologies and software. Coordination can only be done between basic 2D drawings.

BIM设计软件
BIM design software

- BIM技术可实现模型和信息的跨平台转换和传递，最大限度地利用了各专业的设计成果。
- BIM model and information can be delivered between different platforms, allowing full use of results from different disciplines.

C 对比结果
Conclusion

二维设计软件受平台和格式所限，无法实现设计结果和信息传递；而BIM设计软件保证了模型和信息的传递！

The limitations of 2D design software make effective transfer of design results and information between disciplines impossible. With BIM design software, successful flow of design information can be assured.

2.2　BIM技术的八大设计优势
Eight Advantages of BIM Technology

序号 NO.	设计优势 Design Advantage	内容描述 Content Description
1	三维设计 3D Design	项目各部分拆分设计，便于特别复杂项目的方案设计，简单项目的质量优化。 Break up project, convenient for the scheme design of complex project and quality optimization of easy project.
2	可视设计 Visualization Design	室内、室外可视化设计，便于业主决策，减少返工量。 Indoor or outdoor visualization design, convenient for owner to make decision, and reduce the rework.
3	协同设计 Collaboration Design	多个专业在同一平台上设计，实现了高效的协同设计。 Different disciplines work on the same platform; realize a high efficient collaboration design.
4	设计变更 Design alteration	一处修改，处处更新，计算与绘图的融合。 One change, and update in the whole project, to integrate the computation and drawing.
5	碰撞检测 Collision test	通过机电专业的碰撞检测，解决机电管道碰撞。 By collision test in MEP, to eliminate the collision of pipelines.
6	提高质量 Quality improvement	采用阶段协同设计，减少错漏碰缺，提高图纸质量。 Collaboration design in each phase can reduce the mistake and collision, to improve the drawing quality.
7	自动统计 Automatic statistic	可自动统计工程量并生成材料表。 Calculate the quantity statistically, and create material list.
8	节能设计 Energy saving design	支持整个项目绿色节能环保可持续发展。 Support the energy saving, environmental protection and sustainable development of the project.

2.3 BIM技术的八项绿色建筑分析
Eight Analysis of Green Building by Using BIM Technology

八项BIM技术绿色建筑分析

- 建筑能耗模拟分析
  simulation analysis of building energy consumption
- 遮阳与日照模拟分析
  simulation analysis of daylight and sunshade
- 室内、室外热辐射模拟分析
  Simulation analysis of indoor and outdoor heat radiation
- 室内舒适度模拟分析
  Simulation analysis of indoor comfort degree
- 通风效果模拟分析
  Simulation analysis of ventilation
- 建筑光环境分析
  Simulation analysis of luminous environment
- 气候仿真分析
  Simulation analysis of climate
- 地形仿真分析
  Simulation analysis of topography

3

BIM技术标准现状及编制规划

Current Status and Compiling Planning of BIM Technical Standards

3.1 国外与中国香港BIM技术标准现状
Current Status of BIM technical standards Abroad and in Hong Kong

1. 全球BIM发展
Global BIM development

BIM在世界各地（图例）发展迅速，在BIM技术的发展和应用方面，英国、美国、新加坡等国家居于领先地位。

BIM has developed rapidly around the world (as shown on the map), and the UK, US and Singapore are taking the lead in the development and application of BIM technology.

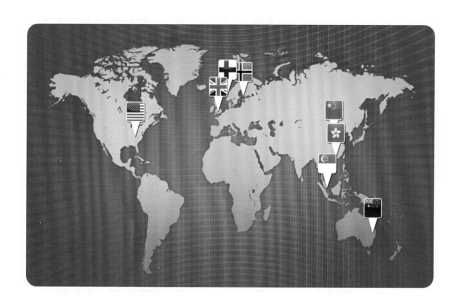

2. 欧美及澳大利亚BIM标准
BIM standards in Europe, America and Australia

> 美国国家BIM标准（NBIMS）
> National BIM standard (NBIMS) of the US

2007年美国国家建筑科学研究院发布了"基于IFC标准制定的BIM应用标准"NBIMS-US (National Building Information Model Standard)的第一个

In 2007, the National Institute of Building Sciences published the first version of "BIM Application Standard Based on IFC Standard" *National Building Information Model Standard* (NBIMS-US). NBIMS-US is an important BIM

67

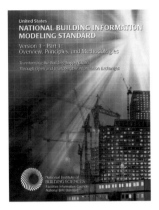

NBIMS-US Ver.1 - 2007

版本。NBIMS-US是一本重要的BIM标准，北美、欧洲、韩国及许多英联邦国家基本上都采纳了美国的BIM标准，或在美国的BIM标准基础上发展自己国家的BIM标准。

standard, and North America, Europe, Korea as well as many Commonwealth countries have adopted American BIM standards, or developed their own BIM standards on the basis of American BIM standards.

> 英国 BIM 标准
UK BIM standards

2010年，英国发布了"AEC（UK）BIM标准"。

In 2010, the UK published *AEC (UK) BIM Standard*.

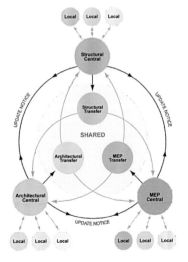

> 挪威 BIM 标准

Norwegian BIM standard

2011年，挪威发布了"BIM手册（1.2版本）"，手册中提出了BIM有关要求和BIM在各个建筑阶段参考用途的信息。

In 2011, Norway published *Statsbygg BIM Manual 1.2*, which provided related requirements for BIM and reference information about BIM at each construction stage.

> 澳大利亚BIM标准

Australian BIM standard

在2009年，CRC Construction Innovation发布了"National Guidelines for Digital Modeling"，面向澳大利亚建筑行业所有参与方的BIM实施指南。

In 2009, CRC Construction Innovation published *National Guidelines for Digital Modeling*, a BIM implementation guide for those in the Australian construction industry.

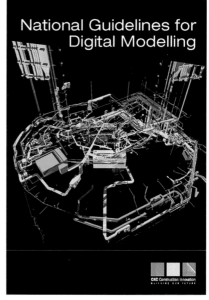

3. 亚洲BIM标准
Asian BIM standards

> 新加坡BIM指导
Singapore BIM Guide

2012年新加坡建设局发布了"BIM指南",包含了BIM规范和BIM建模及协作程序。

In 2012, Singapore's Building and Construction Authority published *Singapore BIM Guide*, which includes BIM specifications, BIM modeling, and collaborative procedures.

> 日本 BIM 标准
Japanese BIM Standard

2012年,JIA(日本建筑学会)发布了"BIM手册"。内容含BIM的技术说明、运作方式、模型建构、结构设计、机电设计、利益、成本费用、附加价值、可能的新价值、成品验证、监造与维护管理。

In 2012, JIA (Japan Institute of Architects) published *BIM Manual*. It contains technical descriptions, modes of operation, modeling, structural design, mechanical and electrical design, benefits, costs, added value, potential new value, finished product verification, supervision, and maintenance management of BIM.

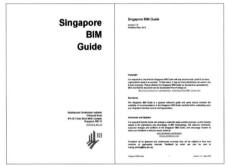

> 韩国 BIM 标准
Korean BIM Standard

2010年，韩国国土海洋部发布了"建筑领域BIM应用指南"，主要是用于指导业主、施工单位和设计师等如何实施BIM技术。

In 2010, Korea's Ministry of Land, Transport and Maritime Affairs published *Guide to BIM Application in Construction Field* mainly to advise owners, contractors and designers to implement BIM technology.

> 中国香港 BIM 标准
Hong Kong BIM Standard

2009年，香港房屋署发布了"建筑信息模拟（BIM）应用标准"。用于制定标准、拟备指引和设立组件资料库。

In 2009, the Hong Kong Housing Authority published "Standard of Building Information Modeling (BIM) Application" to establish standards, prepare guides, and set up a library component.

建筑信息模拟使用指南
Building Information Modeling (BIM) User Guide

建筑信息模拟标准手册
Building Information Modeling (BIM) Standards Manual

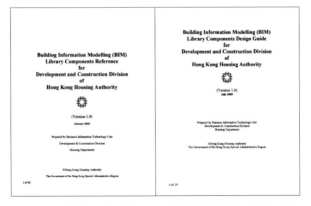

建筑信息模拟组件库参考资料
Building Information Modeling (BIM) Library Components Reference

建筑信息模拟组件库设计指南
Building Information Modeling (BIM) Library Components Design Guide

3.2 中国大陆地区BIM技术标准现状及编制规划
Current Status and Preparation Planning of BIM Technical Standards in the Chinese Mainland

1. 中国BIM框架标准
Chinese BIM framework standard

2011年,清华大学软件学院BIM标准研究课题组发布了《中国建筑信息模型标准框架研究(CBIMS)》,分为理论研究与实证研究两个部分。2012年发布了《设计企业BIM实施标准指南》。

In 2011, *Research of Chinese Building Information Modeling Standard Framework (CBIMS)* was published by the BIM Research Group from the Institute of Software at Tsinghua University, which includes theoretical and empirical research. In 2012, *BIM Implementation Standard of Design Enterprise* was Published.

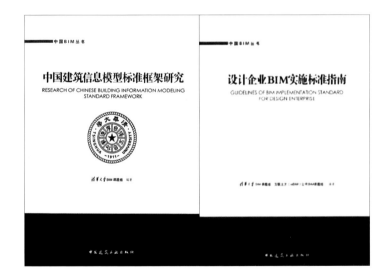

2. 中国BIM行业标准
Chinese BIM industrial standards

2012年,住房与城乡建设部正在研究四项"BIM行业标准"具体如下:

In 2012, the Chinese Ministry of Housing and Urban-Rural Development presented four "BIM industrial standards" as follows:

住房和城乡建设部的四个"BIM行业标准"
Four "BIM Industrial Standards" of the Chinese Ministry of Housing and Urban-Rural Development

序号 SN	项目名称 Name	制订修订 Prepared/ Revised	适用范围和主要技术内容 Applicable scope and main technical content
1	建筑工程信息模型应用统一标准 Uniform Standard of Building Information Modeling (BIM) Application	制订 Prepared	适用于建筑工程全生命期（包括规划、勘察、设计、施工和运行维护各阶段）的信息存储、传递和应用。 主要技术内容：总则，术语和符号，基本规定，规划、勘察、设计、施工、运行维护各阶段的BIM数据及其存储、传递和应用，BIM应用能力评价。 Apply to information storage, transfer, and application during full lifecycle of construction projects (includes all stages, such as planning, survey, design, construction, and operation and maintenance). Main technical content: General rules, terms and symbols, basic requirements, BIM data and its storage, transfer, and application at various stages such as planning, survey, design, construction, and operation and maintenance, assessment of the capability of BIM application.
2	建筑工程信息模型存储标准 Standard of Building Information Modeling (BIM) Storage	制订 Prepared	适用于建筑工程全生命期（包括规划、勘察、设计、施工和运行维护各阶段）模型数据的存储。 主要技术内容：总则，术语和符号，基本规定，BIM数据存储的基本原则、格式要求等。 Apply to storage of model data during full lifecycle of construction projects (including all stages, such as planning, survey, design, construction, and operation and maintenance). Main technical content: General rules, terms and symbols, basic requirements, basic principles of BIM data storage, and format requirements.

主编部门 Editing sector	主编单位 Editing department	参编单位 Department involving preparation	计划完成时间 Completion time in plan
住房和城乡建设部 Chinese Ministry of Housing and Urban-Rural Development	中国建筑科学研究院 China Academy of Building Research	中建三局第一建设工程有限责任公司 清华大学 上海现代建筑设计集团 上海建工集团股份有限公司 The First Construction Engineering Limited Company of China Construction Third Engineering Bureau Tsinghua University Shanghai Xian Dai Architectural Design Group Shanghai Construction Group Co., Ltd.	2013年12月 December 2013
住房和城乡建设部 Chinese Ministry of Housing and Urban-Rural Development	中国建筑科学研究院 China Academy of Building Research	中建三局第一建设工程有限责任公司 清华大学 上海现代建筑设计集团 上海建工集团股份有限公司 The First Construction Engineering Limited Company of China Construction Third Engineering Bureau Tsinghua University Shanghai Xian Dai Architectural Design Group Shanghai Construction Group Co., Ltd.	2013年12月 December 2013

序号 SN	项目名称 Name	制订修订 Prepared/ Revised	适用范围和主要技术内容 Applicable scope and main technical content
3	建筑工程设计信息模型交付标准 Standard of Building Information Modeling (BIM) Delivery	制订 Prepared	适用于建筑工程设计模型数据的交付。 主要技术内容：总则，术语和符号，基本规定，BIM数据交付的基本原则、格式要求、流程等。 Apply to the delivery of BIM data. Main technical content: General rules, terms and symbols, basic requirements, basic principles of BIM data storage, format requirements, and process.
4	建筑工程设计信息模型分类和编码标准 Standard of Building Information Modeling (BIM) Classification and Coding	制订 Prepared	适用于建筑工程设计模型数据的分类和编码。 主要技术内容：总则，术语和符号，基本规定，BIM数据分类和编码的基本原则、格式要求等。 Apply to classification and BIM data coding. Main technical content: General rules, terms and symbols, basic requirements, basic principles of BIM data classification and coding, and format requirements.

（续表）

主编部门 Editing sector	主编单位 Editing department	参编单位 Department involving preparation	计划完成时间 Completion time in plan
住房和城乡建设部 Chinese Ministry of Housing and Urban-Rural Development	中国建筑标准设计研究院 China Institute of Building Standard Design & Research	中国建筑设计研究院 北京市建筑设计研究院 上海现代建筑设计（集团）有限公司 清华大学 上海交通大学 中建国际设计顾问有限公司 中国建筑工程总公司 北京金土木软件技术有限公司 北京理正软件设计院有限公司 China Architecture Design & Research Group Beijing Institute of Architectural Design Shanghai Xian Dai Architectural Design (Group) Co., Ltd. Tsinghua University Shanghai Jiao Tong University China Construction Design International Co. Ltd. China State Construction Euginering Corporation Beijing Civil King Software Technology Co., Ltd. Beijing Leading Software Design Institute, Co., Ltd.	2013年12月 December 2013
住房和城乡建设部 Chinese Ministry of Housing and Urban-Rural Development	中国建筑标准设计研究院 China Institute of Building Standard Design & Research	中国建筑设计研究院 北京建筑材料科学研究总院 中科院建筑设计院有限公司 上海现代建筑设计（集团）有限公司 China Architecture Design & Research Group Beijing Building Materials Academy of Sciences Research Institute of Architecture Design and Research, Chinese Academy of Sciences Shanghai Xian Dai Architectural Design (Group) Co., Ltd.	2013年12月 December 2013

3.3 BIM技术标准简述
Brief Introduction of BIM Technical Standards

1. BIM标准的层级划分
Division levels of BIM standards

按照逻辑结构层级划分,可分为四级(如下图):

BIM technical standards can be divided into four levels based on a logical structure (as shown in the following):

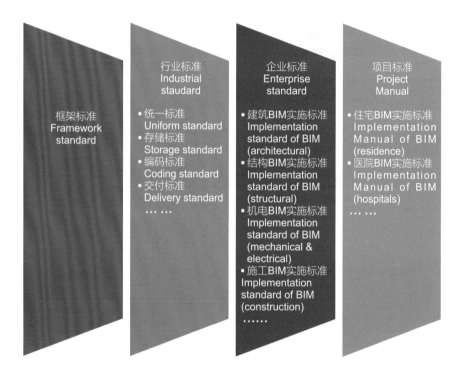

1) 框架标准:主要设定了BIM标准的理论基础

相关标准:《中国建筑信息模型标准框架研究》(CBIMS);《设计企业BIM实施标准指南》

1) Framework standard: establish theoretical foundation of BIM standards

Related standards: *Research of Chinese Building Information Modeling Standard Framework (CBIMS); BIM Implementation Standard of Design Enterprise*

2）行业标准：设定行业内BIM应用相关方面的统一标准；数据交互、调用存储等方面的基础标准以及依据与软件平台的执行标准。

相关四个标准：
《建筑工程信息模型应用统一标准》
《建筑工程信息模型存储标准》
《建筑工程设计信息模型交付标准》
《建筑工程设计信息模型分类和编码标准》

3）企业标准：结合企业自身特点、应用范围及应用平台制定的具有一定针对性的企业内部标准。

相关标准：
《中国建筑设计研究院BIM实施手册》

4）项目标准（手册）：针对特定项目类型制定的BIM标准。

2) Industrial standard: Establish uniform standards for BIM application in the industry, basic standards for data exchange, coding and storage, and basic standards for software platforms

Four related standards:
Uniform Standard for Application of Building Information Modeling (BIM)
Storage Standard of Building Information Modeling (BIM)
Standard of Building Information Modeling (BIM) Delivery
Standard of Building Information Modeling (BIM) Classification and Coding

3) Enterprise standard: Enterprises' own standards are specifically developed based on the features, application scope, and application platform.

Related standards:
BIM Implementation Manual of China Architecture Design & Research Group

4) Project standard: BIM standards developed based on specific project types.

2. BIM标准的编制基础
Basis for BIM standard preparation

1）编制思路：BIM技术、BIM标准、BIM软件的同步发展。

以中国建筑工程专业应用软件与BIM技术紧密结合为基础，首先开展专业BIM技术和标准的课题研究，用BIM技术和方法改造专业软件，形成专业BIM，将专业BIM集成，形成阶段BIM，最后将各阶段BIM连通，形成项目全生命期BIM。

在框架BIM课题成果（技术和软件）达到可实施的基础上，按照BIM标准的结构层级制定其他各层次的BIM标准。

2）软件平台：目前针对BIM的软件平台较多，但都具有以下特征：

1) Preparation ideas: synchronized development of BIM technology, standards, and software.

Based on the close integration of professional application software in Chinese construction projects and BIM-based technology, firstly conduct subject research on professional BIM technologies and standards, transform professional software by using BIM technologies and methods to form professional BIM. Then integrate the professional BIM to form various BIM stages, and finally connect each BIM stage to get a BIM formulate the full lifecycle of the project.

When the achievements (in technology and software) of the BIM framework research can be implemented, other levels of BIM standards can be developed according to the BIM standard structural level.

2) Software platform: Currently there are many software platforms for BIM, and all have the following features:

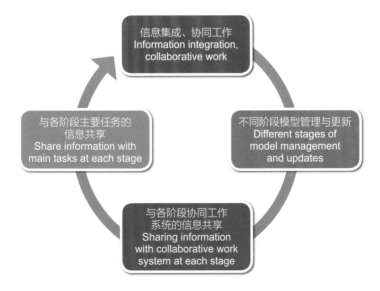

3）编制范围：BIM标准的编制一般主要涉及以下三个方面，三者缺一不可：

3) Scope of preparation: Generally preparation of BIM standards involve the following three aspects, all of which are indispensable:

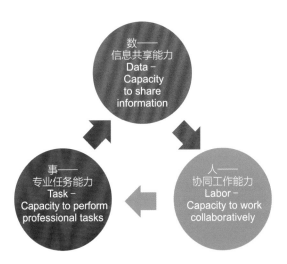

a）数——信息共享能力，是BIM的核心（IT：信息内容、格式、交换、集成和存储）

b）人——协同工作能力，是BIM的应用过程（管理：流程优化、辅助决策，与传统方式的不同）

c）事——专业任务能力，是BIM的目标（专业：专业BIM软件、专业标准、完成专业任务的效率和效果）

a) Data – Capacity to share information, core of BIM (IT: information, format, exchange, integration, and storage)

b) Labor – Capacity to work collaboratively (Management: process optimization, assist in decision-making, different from the traditional way)

c) Task – Capacity to perform professional tasks, objective of BIM (Professional: professional BIM software, professional standards, and efficiency and result of completing professional tasks)

参考文献

1. 2007年美国国家建筑科学研究院发布的"基于IFC标准制定的BIM应用标准";

2. 2009年CRC Construction Innovation发布的"National Guidelines for Digital Modeling";

3. 2010年英国发布的"AEC（UK）BIM标准（专为Autodesk Revit）";

4. 2011年挪威发布的"BIM手册版本1.2";

5. 2012年新加坡建设局发布的"BIM指南";

6. 2012年JIA（日本建筑学会）发布的"BIM手册";

7. 2010年韩国国土海洋部发布的"建筑领域BIM应用指南";

8. 2009年香港房屋署发布了"建筑信息模拟（BIM）应用标准";

9. 2011年清华大学BIM标准研究课题组发布的"中国建筑信息模型标准框架研究"（CBIMS）和《设计企业BIM实施标准指南》。

References

1. "BIM Application Standard Based on IFC Standard" published by the US National Institute of Building Sciences in 2007;

2. "National Guidelines for Digital Modeling" published by CRC Construction Innovation in 2009;

3. "AEC (UK) BIM Standard (for Autodesk Revit)" published by the UK in 2010;

4. Statsbygg BIM Manual 1.2 published by Norway in 2011;

5. Singapore BIM Guide published by Singapore's Building and Construction Authority in 2012;

6. BIM Manual published by JIA (Japan Institute of Architects) in 2012;

7. Guide to BIM Application in Construction Field published by Korea's Ministry of Land, Transport and Maritime Affairs in 2010;

8. "Standard for Building Information Modeling (BIM) Application" published by the Hong Kong Housing Authority in 2009;

9. Research of Chinese Building Information Modeling Standard Framework (CBIMS) and BIM Implementation Standard of design enterprise published by the BIM Research Group from the Institute of Software at Tsinghua University in 2010.

4

BIM技术在企业应用方面的推广

BIM Promotion in Enterprise Applications

4.1 BIM 设计技术发展模式
Development Model of BIM Design Technology

模式 Mode	定义 Definition	主要特点及优势 Major Features & Advantages	主要问题及不足 Major Problems & Limitations
BIM设计模式 BIM Design mode	将BIM设计技术定位为符合企业发展战略、技术革新和核心竞争能力增强的重要技术及信息化手段；主动将BIM技术应用于企业主流设计业务中；通过将BIM工作流程替代传统二维流程实施，以此全面提升设计质量和设计团队的BIM全面应用能力。 BIM design technology is positioned as an important means of technology and information that coincides with the enterprises' development strategy, technological innovation, and enhanced core competitiveness; actively adopt BIM in enterprises' mainstream design business; replace the conventional 2D process with the BIM workflow to comprehensively improve design quality and the ability of design teams to use BIM.	1.设计师直接在BIM平台上设计； 2.全过程、全方位、实时设计优化与协调； 3.精细化设计提高整体设计质量； 4.节约项目成本和投资，缩短项目周期； 5.全面提高企业竞争力。 1. Designers directly design on the BIM platform; 2. Whole process, comprehensive, and real-time design optimization and coordination; 3. Detailed design to improve overall design quality; 4. Save project cost and investment, reduce project time period; 5. Comprehensively increase enterprises' competitiveness.	1. 投入较大，见效较慢； 2. 涉及思维、专业、流程的范围较大，因此推广阻力较大； 3.缺乏合理的与工作量相匹配的分配及奖励机制。 1. Requires tremendous investment, and the effect is slower; 2. Involves a larger range of thinking, professions, and processes, and so it faces big obstacles during promotion; 3. The lack of distribution matches the workload of the reasonable and Incentive mechanism.

（续表）

模式 Mode	定义 Definition	主要特点及优势 Major Features & Advantages	主要问题及不足 Major Problems & Limitations
BIM验证模式 BIM verification mode	由于设计市场和技术手段的需要，被动将BIM技术应用于企业内部分项目中；通过采用传统二维设计流程和BIM工作流程两条线并行实施，以BIM验证传统二维工作成果，以此提升项目设计质量。 Demanding of the design market and technological means results in the passive application of BIM in some enterprises' internal projects; by concurrently implementing the traditional 2D design process and BIM workflow, use BIM to verify work outcomes of traditional 2D, thus improving projects' design quality.	1.设计师用传统二维设计后，用BIM技术进行成果验证； 2.阶段性设计优化与协调； 3.局部完善和提高设计质量； 4.局部节约项目成本和投资，缩短项目周期； 5.投资较小，见效较快。 1. Designers use BIM to verify the outcome after designing with conventional 2D; 2. Phase in design optimization and coordination; 3. Locally perfect and improve design quality; 4. Locally save project costs and investment, reduce project time period; 5. Require smaller investment, and quick results.	1.和主流设计业务脱节，不利于企业竞争力的提高； 2.设计成果多次转译造成信息损失； 3.重复劳动造成设计成本的增加。 1. Disconnected from mainstream design business, unfavorable for improving enterprises' competitiveness; 2. Multi-translations of the design outcome cause information to be lost; 3. Duplication caused an increase in design cost.

结论：以上两种模式各有特点，需要企业决策者根据自身情况选择合适的模式。从行业、企业的健康持续发展的角度考虑，我们建议采用"BIM设计模式"。

Conclusion: The above two modes have their own features, and the enterprises' decision makers should select the most appropriate one based on their own situation. For the healthy and continuous development of industry and enterprises, we recommend "BIM design mode".

4.2 中国建筑设计研究院 BIM 技术发展历程
Development History of BIM in CAG

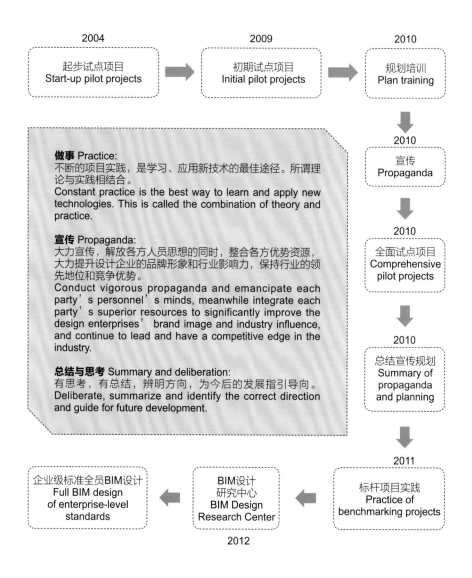

4.3 中国院在企业BIM方面推广的十一个步骤
Eleven Promotion Steps of BIM in Enterprises by CAG

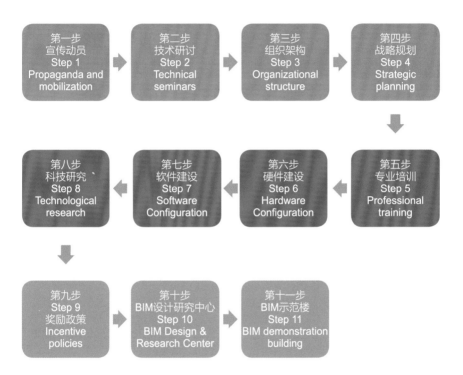

1. 第一步： 宣传动员

Step 1: Propaganda and mobilization

企业决策层至一线员工对BIM技术高度及持续关注，转变传统的思维模式，是企业推广应用BIM技术的先决条件和有力保障。

Enterprise decision-makers to frontline employees must pay high and consistent attention to BIM technology and change their traditional way of thinking, which is the prerequisite and strong guarantee for enterprises to promote and implement BIM technology.

集团院长、总建筑师、总会计师、总工程师等集团领导到场参会
Leaders of the enterprise group, such as the president, chief architects, chief accountants, and chief engineers attended the meeting

1）2010年11月成功举办了"中国院第一届BIM设计研讨会（2010年）"（动员大会）

会议主题："改变建筑设计的手段，提高企业核心竞争力"

参会人员：计划150人，实际到会设计人员400人。

1) In November 2010, the 1st CAG Seminar on BIM Design (2010) (mobilization meeting) was successfully held

Subject: "Changing architectural design method to improve the enterprise's core competitiveness"

Number of participants: While 150 participants were expected, in fact 400 designers attended the meeting.

会议场面：集团领导高度重视，设计人员踊跃参加，十年久违的热烈场面。

2）2011年11月成功举办了"中国院第二届BIM设计研讨会"

会议主题："BIM设计项目推广及研讨"

参会人员：460位设计人员

Scene of the meeting: With special attention from the group's leaders and designers' active participation, it was a warm scene to be remembered for many years.

2) In November 2011, the 2nd CAG Seminar on BIM Design was successfully held

Subject: "The promotion and discussion of BIM design projects"

Participants: 460 designers

3）2012年11月成功举办了"中国院第三届BIM设计研讨会"

会议主题："BIM标准、BIM市场和BIM实践"

参会人员：建设单位、设计单位、软件厂商等300多名代表参加了会议。

3) In November 2012, the 3rd CAG Seminar on BIM Design was successfully held

Subject: "BIM standards, BIM market, and BIM practice"

Participants: More than 300 delegates from owners, design units and software manufacturers attended the meeting.

4）出版了《中国建筑设计研究院2010年BIM项目设计研讨会资料汇编》，编写了《中国院BIM大事记》等。

4) Materials such as *Compilation of 2010 CAG Seminar on BIM Project Design* and *CAG BIM Memorabilia* were compiled and published.

5）中国院BIM培训材料汇编及报刊宣传。

5) CAG's compendium of BIM training materials and newspaper articles.

2. 第二步：技术研讨
Step 2: Technical seminars

1) 组织 "BIM行业专家技术交流会（2011）"

1) Organize "Technical exchange meeting for BIM experts" (2011)

特邀嘉宾
中国建筑设计研究院（集团）院长助理
欧阳东
北京市建筑设计研究院执行总建筑师
邵韦平
中建国际公司副总监兼业务部经理
过 俊
上海现代建筑设计集团BIM工作组组长
徐 浩
昆明市建筑设计研究院副总建筑师
许 峻
清华大学建筑设计研究院BIM中心主任
陈宇军
中国建筑设计研究院国际所所长
魏篙川

Special guests
Ouyang Dong, Assistant President of CAG (Group)
Shao Weiping, Executive Chief Architect of Beijing Institute of Architectural Design
Guo Jun, Deputy Director & Business Department Manager of China State Construction International Company
Xu Hao, Leader of BIM Working Group of Shanghai Xian Dai Architectural Design Group
Xu Jun, Deputy Chief Architect of Kunming Architectural Design & Research Institute
Chen Yujun, Director of BIM Center of Architectural Design and Research Institute of Tsinghua University
Wei Gaochuan, President of the International Institute of China Architecture Design & Research Group

座谈主题：各单位项目应用BIM的情况，使用BIM遇到的最大问题，BIM项目的工作流程（方案—扩初—施工图），项目比例推广模式和分配方式，全专业协同时的工作模式等。

Topic: Implementation of BIM in each unit's projects, the biggest obstacles encountered during BIM usage, workflow of BIM projects (plan - development design - working drawings), promotional mode and distribution method of project scale, working mode of full-range specialty coordination.

2）中国院与知名BIM厂商进行高层交流

2) CAG conducted high-level communications with well-known BIM manufacturers

中国院的各级领导与国外知名厂商探讨BIM技术的合作方式。
CAG leaders at all levels discuss with renowned foreign manufacturers on cooperation modes for BIM technology.

3. 第三步：组织架构
Step 3: Organizational Structure

1）成立中国院BIM组织机构——BIM技术委员会和BIM工作小组

1) CAG established a BIM organization–BIM technical committee and BIM working group

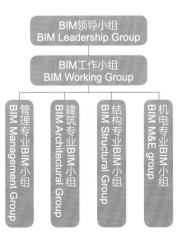

2) BIM工作小组的主要任务
Main tasks of BIM working groups

专业 小组 Group	建筑专业 BIM组 BIM Architectural Group	结构专业 BIM组 BIM Structural Group	机电专业 BIM组 BIM M&E Group	管理专业 BIM组 BIM Management Group
组长（教授级高建（工）担任） Group leader (professor-level senior architect/engineer)				
副组长（BIM技术骨干）担任 Deputy group leader (BIM technical backbone)				
成员 Group member				
主要任务 Main tasks	1）本专业的BIM手册编制； 2）各专业共同形成的工作设计流程； 3）BIM项目信息管理； 4）BIM技术指导，编制答疑文件； 5）各专业的工作界面； 6）BIM相关课题； 7）族文件的归档和复用； 8）与现有二维平台的对接。 1) Prepare BIM manuals of each discipline 2) All disciplines work together to form the design process; 3) Manage BIM project information; 4) BIM technical guidance, prepare Q&A documents; 5) Work interfaces of each specialty; 6) BIM related subjects; 7) Archive and recycle family documents; 8) Connect with existing 2D platforms.			1）参与中国院的"五年BIM战略规划"的制定； 2）对外咨询的协调； 3）组织BIM技术培训； 4）编制服务分包合同； 5）相关文件资料管理； 6）会议活动组织。 1) Participate in preparing CAG's "Five-Year Strategic Planning for BIM"; 2) Coordinate external consulting; 3) Organize BIM technological training; 4) Prepare service subcontracts; 5) Manage related documents; 6) Organize meetings.

4. 第四步: 战略规划
Step 4: Strategic Planning

制定了《中国院BIM五年战略规划》

Prepared CAG Five-Year Strategic Planning for BIM

《中国院BIM五年战略规划》目录

1 BIM能给中国院带来什么
　1.1 总体目标
　1.2 中国院的现状及特点
　1.3 中国院BIM方案的独特模式
　1.4 本解决方案主要完成的内容
　1.5 本解决方案实施后的效果
2 针对中国院的BIM解决方案体系
　2.1 中国院BIM解决方案总揽
　2.2 中国院BIM解决方案简介
3 本解决方案实施
　3.1 实施的思路和要求
　3.2 BIM年度实施计划
　3.3 实施的组织构成
4 知识产权
5 预算
6 附录
附录一: 培训
附录二: 技术支持
附录三: 二次开发

Contents of CAG Five-Year Strategic Planning for BIM

1 What will BIM Bring to CAG
　1.1 Overall Objectives
　1.2 Current Status and Features of CAG
　1.3 Unique Mode of CAG's BIM Solution
　1.4 What will be Completed with this Solution
　1.5 Effect of Implementing this Solution
2 BIM Solution System for CAG
　2.1 Overview of CAG's BIM Solution
　2.2 Brief Introduction of CAG's BIM Solution
3 Implementing this Solution
　3.1 Ideas and Requirements of Implementation
　3.2 Annual BIM Implementation Plan
　3.3 Organization of Implementation
4 Intellectual Property
5 Budget
6 Annexations
Annexation 1: Training
Annexation 2: Technical Support
Annexation 3: Secondary Development

5. 第五步：专业培训
Step 5: Professional training

专业的BIM基础技术培训是后续BIM项目试点和全面推广应用的重要技术保障手段。

1）进行了中国院BIM全专业基础技术培训（6批200多个设计师）

2）中国院BIM基础技术培训的结业证书、奖励证书（含奖金）

3）中国院"BIM学习申请及审批表"及"BIM学习日志"

Professional training on BIM basic technology is an important means of technical support for subsequent BIM pilot projects and full promotion and implementation.

1) Provided CAG full-range specialty training on BIM basic technology (for more than 200 designers in six groups)

2) Certificates of achievement and awards (including bonuses) of CAG training on BIM basic technology

3) CAG "Application and Approval Forms for BIM Learning" and "BIM Learning Logs"

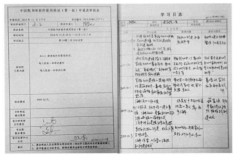

6. 第六步：硬件建设
Step 6: Hardware Configuration

1）建设中国院信息网络中心机房

网络实现万兆主干、千兆到桌面；服务器、UPS电源、精密空调均采用最先进的产品。满足五年信息化发展规划，核心交换机房采用数据双热备份。

2）配置计算机终端设备

由于全专业设计采用BIM配置要求较高，所以购置了运算速度较快的计算机终端设备。

1) Build computer room for CAG's Information and Network Center

The network is gigabit backboned and GTTD (Gigabit to the Desktop); all servers, UPS power supplies, and precision air conditioners are the most advanced. These can meet the five-year development plan for information, with dual hot backup of data installed in the core switch room.

2) Configure computer terminals

Since the full-range specialty design uses BIM and has a high demand for configuration, computer terminals with high-speed processors were purchased configured.

7. 第七步： 软件建设
Step 7: Software Configuration

中国院购买使用的BIM设计软件　　　　　CAG purchased and implemented BIM design software

 建筑设计 Architectural design

 结构设计 Structural design

 给水排水、暖通空调、电气设计 Plumbing, HAVC, and electrical design

MEP 碰撞检测、虚拟漫游 MEP collision detection, virtual roaming

日照分析、能耗分析等 Sunlight analysis and energy consumption analysis

制造前的可视化和仿真 Visualization and simulation before production

 给水排水、暖通空调、电气设计 Plumbing, HAVC, electrical design

方案设计模拟环境 Scheme design simulation environment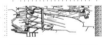

 模拟真实光环境下的建筑观感 Building appearance under simulated real light conditions

 建筑节能分析 Analysis of building energy efficiency

8. 第八步：课题研究
Step 8: Subject Research

开展了中国院、北京市、住宅和城乡建设部的BIM科研课题研究

CAG, Beijing, and the Chinese Ministry of Housing and Urban-Rural Development conducted subject researches on BIM.

	企业（中国院）课题 Research subject of Enterprise (CAG)	北京市课题 Research subject of Beijing	住房和城乡建设部课题 Research subject of the Chinese Ministry of Housing and Urban-Rural Development
名称 Name	《BIM在建筑工程设计中的应用技术研究》 *Technical Research on BIM Application in the Design of Construction Projects*	《BIM技术在龙岩金融商务中心（B地块）B3#楼项目中的应用研究》 *Research on BIM Application in Building B3# Project of Longyan Financial Business Centre (Plot B)*	《中国民用建筑BIM应用方法论与实施方案研究》 *Research on BIM Application Methodology and Implementation Plan on China's Civil Buildings*
合作单位 Cooperation Unit	无 None	无 None	清华大学；欧特克公司 Tsinghua University; Autodesk Inc

（续表）

	企业（中国院）课题 Research subject of Enterprise (CAG)	北京市课题 Research subject of Beijing	住房和城乡建设部课题 Research subject of the Chinese Ministry of Housing and Urban-Rural Development
成果 Outcome	1）"中国院BIM战略规划报告"； 2）"中国院建筑专业BIM手册"； 3）"中国院结构专业BIM手册"； 4）"中国院机电专业BIM手册"； 5）"BIM示范工程设计案例报告"。 1) "Report of CAG BIM Strategic Planning"； 2) "CAG BIM Architectural Manual"； 3) "CAG BIM Structural Manual"； 4) "CAG BIM M&E Manual"； 5) "Report of BIM Design Cases for Demonstration Projects".	1）"建筑、结构、给排水、暖通空调专业、电气、电信专业BIM设计图纸"； 2）"建筑模板文件（初步版本）"； 3）"各专业（结构和机电）含有技术参数BIM模型文件"； 4）"软件数据交互插件（内部试用版）"； 5）"中国建筑设计研究院BIM实施手册"。 1) "BIM Design Drawings of Architectural, Structural, Plumbing, HVAC, Electrical, and Communication Specialties"； 2) "Building Template Files (preliminary version)"； 3) "Technical Parameters Contained in BIM Model Documents for Structural and M&E Specialties"； 4) "Plug-ins for Software Data Exchange (internal trial version)"； 5) "CAG BIM Implementation Manual".	1）"中国民用建筑BIM应用方法论"； 2）"民用建筑BIM实施研究报告"； 3）"民用建筑BIM建筑手册"。 1) "Methodology of BIM Application on Chinese Civil Buildings"； 2) "Research Report on BIM Implementation on Civil Buildings"； 3) "BIM Construction Manual for Civil Buildings".

9. 第九步：激励措施
Step 9: Incentive policies

1）奖励政策：制定了"采用BIM技术进行全专业设计的奖励政策"；全专业使用BIM设计，奖励总设计费（扣除分包）的5%～8%；设计收费为400万及以上时，奖励5%。

2）项目评估：制定了"中国院BIM设计可行性分析表"

1) Reward policy: Prepared "Reward policy for implementing BIM in full-discipline design", for designs totally adopting BIM, 5%～8% of the total design fee will be awarded (deducting subcontracted parts); and 5% will be awarded when the design fee is no less than RMB 4 million.

2) Project evaluation: Prepared "CAG Feasibility Analysis Table of BIM Design"

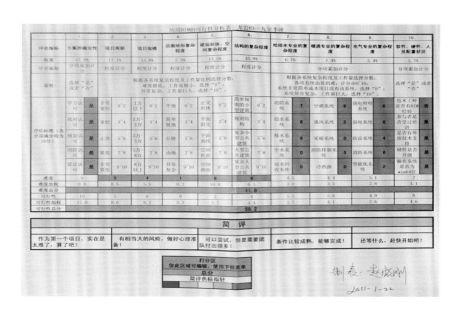

3）项目补助：制定了"中国院BIM设计补助申请及审批表"。

3) Project subsidy: Prepared "CAG Application and Approval Form for BIM Design Subsidy".

BIM设计补助申请及审批表
An Approval Application Form for BIM Desigh Subsidy

申请时间：2011年 月 日　　　　　　BIM设计编号：2011-BIM（　）
Application date:　　　　　　　　　　BIM design No.: 2011-BIM（　）

申请部门 Application department		项目经理 Project manager					
项目名称 Project name							
设计号 Design No.		总设计费/分包设计费 Total desigh fee / Subcontrating design fee		元 RMB /		元 RMB	
项目性质 Nature of the project		建筑面积 Floor area					
BIM设计主持人及工种负责人 BIM design director and disciptiue principal							
专业 Specialty	建筑 Architectural	结构 Structural	水 Plumbing	暖 HVAC	电 Electrical	智能化 Building Intelligence	总图 Master Design
室名称 Studio name							
分配比例 Distribution ratio							
参与BIM设计的研究室主任意见 Opinions of research director participating in BIM design							

（续表）

BIM设计的主要成果 The main results of BIM design	建筑专业：BIM 出图率100%（部分三维图中的二维修饰） 100% BIM architectural drawing output (some drawings may be modified in 2D); 结构专业：BIM 出图率70%（部分三维图中的二维修饰） 70% BIM structural drawing output (some drawings may be modified in 2D); 机电专业：BIM 出图率70%（部分三维图中的二维修饰） 70% BIM M&E drawing output (some drawings may be modified in 2D); MEP collision detection; architectural animation. 机电MEP碰撞检查。建筑动画。 MEP collision detection; architectural animation.
BIM补助款 BIM subsidy	（总设计费-分包设计费）*（5-8）% = (the total design fee - subcontracting design fee) *（5-8）% =
批准补助预付款 Approval for advance payment	万元（BIM补助费用的40%），拨付时间为批准日后十个工作日。 Million yuan (BIM subsidy costs 40%), allocate time for approval within ten working days.
剩余补助费用（完成施工图后） Remaining subsidy	万元（项目施工图完成后十天内，由院（集团）付） Million yuan (ten days after the completion of project construction, to be paid by CAG 批准的时间：　　　年　　　月　　　日 Date of approval:　　　Year　　　Month　　　Day
总监或名人工作室主任意见 Opinions of project director or studio director	
专业院院长意见 Opinions of institute president of each specialty	
财务部负责人意见 Opinions of principal of Financial Department	
设计运营中心主任 Director of design operations center	
集团领导 CAG leader	

4）项目管理：通过"BIM设计实施手册"和"中国院BIM会议纪要"等进行过程管理。

4) Project management: Perform process management by using "BIM Design Implementation Manual" and "CAG BIM Meeting Minutes".

BIM资源标准　BIM Resource Standard
　计算机软硬件配置标准
　Computer software and hardware configuration standards
　BIM资源库建立标准
　BIM resource library to establish standards
BIM行为标准　BIM standards of conduct
　模型搭建规则　Modeling rules
　模型内容深度标准　Standards of model depth
　命名规则　Naming rules
　BIM制图标准研究
　Research on BIM drawing standards
BIM交付标准　BIM delivery standards
　业主、设计单位、施工单位信息模型交付标准
　Information model delivery standards from owners, design units, and construction units
　不同软件间数据交互标准
　Standard of data exchange between different software
　归档文件标准　Standards of archived documents

5）项目验收：中国院BIM设计验收表

5) Project acceptance: CAG BIM Design Acceptance Form

BIM设计补助余额申请审批及验收表 An Application and Acceptance Form for Remaining BIM Design Sudsidy							
申请时间： 年 月 日 Application date:				BIM设计编号：2011-BIM（ ） BIM design No.: 2011-BIM（ ）			
申请部门 Application department		项目经理 Project manager					
项目名称 Project name							
设计号 Design No.		已补助总金额 Total amount of subsidy already paid		万元 RMB			
项目性质 Nature of the project		建筑面积 Floor area					
BIM设计主持人及工种负责人 BIM design director and discipline principal							
专业 Specialty	建筑 Architectural	结构 Structural	水 Plumbing	暖 HVAC	电 Electrical	智能化 Building Intelligence	总图 Master Design
室名称 Studio name							
分配比例 Distribution ratio							
参与BIM设计的研究室主任意见 Opinions of research director participating in BIM design							

（续表）

		完成情况确认（BIM工作小组确认签字）Confirmation of completion (Signature of BIM working group for confirmation)
BIM项目验收 BIM project acceptance	验收指标：【建筑专业BIM 出图率100%（部分图纸可二维修饰）结构专业BIM 出图率70%（部分图纸可二维修饰）；机电专业BIM 出图率70%（部分图纸可二维修饰）；机电MEP碰撞检查；建筑动画】。 Acceptance Index: [100% BIM architectural drawing output (some drawings may be modified in 2D); 70% BIM structural drawing output (some drawings may be modified in 2D); 70% BIM M&E drawing output (some drawings may be modified in 2D); MEP collision detection; architectural animation].	
	BIM图纸A3册，各专业可以合订也可分别成册； A3 catalogues of BIM drawings, the drawings of each specialty can be bound together or separately;	
	碰撞检测报告（含首次碰撞检测点和调整碰撞点后的检测报告） Reports on M&E collision detection (including detection reports on the first collision detection detention and after collision point adjustment)	
	1~3分钟的模型漫游，含全专业模型； Model roaming of 1~3 minutes, including all specialty models;	
	全专业BIM模型完成情况； Completion BIM models of all disciplines	
	完成BIM设计相关电子文件的归档 Complete archive of BIM design related electronic documents	
剩余补助费用 Remaining subsidy	元 批准的时间： 年 月 日 RMB Date of approval:	
总监或名人工作室主任意见 Opinions of project director or studio director		
专业院院长意见 Opinions of institute president of each specialty		
财务部负责人意见 Opinions of principal of Financial Department		
设计运营中心主任 Director of Design Operations Center		
集团领导 CAG leader		

10. 第十步：BIM研究中心
Step 10: BIM Research Center

通过几年的BIM技术、BIM经验及BIM人才积累，BIM发展已达到一定水平，为了更加规范化推进企业BIM发展。2012年8月成立了"中国建筑设计研究院（集团）建筑设计总院BIM设计研究中心"。

BIM设计研究中心的主要职能：
1）承接并完成各类BIM设计项目
2）研究BIM设计措施及标准
3）开拓BIM的商业模式
4）企业内部BIM技术支撑及推广

After several years' accumulation of BIM technology, experience and talents, BIM has reached a certain level. To further promote standard BIM development in CAG, the "Research Center of BIM Design of General Institute of Architecture Design of CAG (Group)" was founded in August 2012.

Main functions of the BIM Research Center:
1) Undertake and accomplish various BIM design projects;
2) Research BIM design measures and standards;
3) Expand BIM business model;
4) Offer internal BIM technical support and promotion to CAG.

11. 第十一步: BIM科研楼
Step 11: BIM Scientific Research Building

2013年全专业用BIM-设计"中国院BIM科研示范楼"

In 2013, all specialties implemented BIM to design the "CAG BIM Scientific Research Building"

工程概况：建筑面积约3.5万m^2、建筑高度60m；绿色三星；智能停车。
Project overview: floor area: 35,000m^2; building height: 60m; Three-star Green design; smart parking.

4.4 参加行业BIM专题讲演推广BIM技术
Promoting BIM by Participating in Industry Special Lectures on BIM

1. 参加了"欧特克AU中国'大师汇'(2011)"(北京)
Participated in "Autodesk University (AU) 2011" (Beijing)

中国院(集团)院长助理欧阳东在2011年11月"欧特克AU中国'大师汇'(2011)"(北京)作题为"构建以BIM为核心力的中国建筑设计企业"的主旨发言,反响非常热烈。来自中国内地、台湾、香港及澳门的各行业的设计单位、施工单位、建设单位、大学及科研单位等1400多位代表参加了本次盛会。

At the "Autodesk University (AU) 2011" (Beijing), Ouyang Dong, Assistant President of CAG (Group) gave a keynote speech entitled, *Building a Chinese Design Enterprise with BIM as Core Competitiveness*, which received an enthusiastic response. More than 1,400 delegates from design units, construction units, owners, universities, and scientific research institutes from the Chinese Mainland, Taiwan, Hong Kong, Macao attended this grand event.

2. 参加了"中央企业CIO年会（2012）"（长沙）
Participated in CIO of Central SOE Annual Conference 2012 (Changsha)

"BIM设计研究中心"于洁主任在2012年11月"中央企业CIO年会"（长沙）会上作专题发言，受到国资委领导的赞扬。国务院国资委及有关司局负责人，各中央企业信息化、科技部门代表共约600人参加会议。

Yu Jie, Director of CAG BIM Research Center gave a keynote speech during the CIO of Central SOE Annual Conference (Changsha) held in November 2012 and received praise from leaders of the State-owned Assets Supervision and Administration Commission of the State Council (SASAC). About 600 participants such as principals of SASAC and related departments and bureaus, delegates from information and technological departments of various central enterprises attended the meeting.

3. 中国-新加坡工程建设行业建筑信息模型（BIM） 高峰交流会（2012）（新加坡）
Sino-Singapore Building Information Modeling (BIM) Executive Summit (2012) (Singapore)

中国院（集团）院长助理欧阳东在2012年11月 "中国-新加坡工程建设行业建筑信息模型（BIM）高峰交流会（2012）"（新加坡）作题为"未来的建筑设计革命-BIM技术"的主题发言。来自中国（含中国内地、台湾、香港及澳门）和新加坡的100多位BIM国际专家和嘉宾参加了本次会议。

During the Sino-Singapore Building Information Modeling (BIM) Executive Summit (2012) held in November 2012, Ouyang Dong, Assistant President of CAG (Group) gave a keynote speech entitled, *Future Building Design Revolution – BIM Technology*. More than 100 international experts in BIM and guests from China (including the Chinese Mainland, Taiwan, Hong Kong, and Macao) and Singapore attended this meeting.

4.5　BIM技术设计培训的八个步骤
Eight Steps of BIM Technology Design Training

BIM技术培训分为以下8大步骤：
BIM technology training includes the following 8 steps:

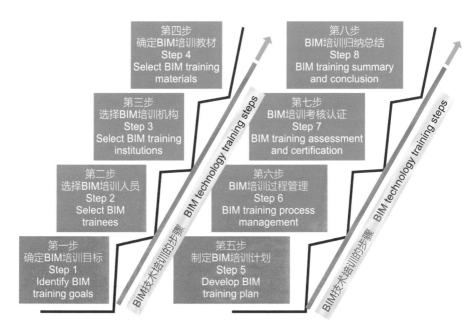

1. 第一步：确定BIM培训目标
Step 1: Identify BIM training goals

层级 Level	培训目标 Training goals
企业级BIM培训 Enterprise-level BIM training	主要针对BIM对行业及企业发展战略、设计竞标、设计流程、质量控制及数据管理等的提升。 Focused on using BIM to improve industry and enterprise development strategy, design bidding, design flow, quality control, and data management.
项目级BIM培训 Project-level BIM training	主要针对BIM技术软件基本技能、高级应用技巧的提升 Focused on using BIM technical software to improve basic skills and advanced application skills

注：不同层级的BIM培训目标，将决定不同的培训内容、培训计划和培训结果。

Note: BIM training goals of different levels will have different training content, plans and results.

企业级BIM培训目标
Enterprise-level training goals

- BIM技术对行业、企业未来发展价值宣讲
- BIM对项目竞标经营价值宣讲
- BIM对业务流程、质量控制、数据管理价值宣讲
- Promote BIM's value in the industry and enterprises' future development
- Promote BIM's value in project bidding and operation
- Promote BIM's value in business process, quality control, and data management

项目级BIM培训目标
Project-level BIM training goals

- 全员BIM基本应用技能普及
- BIM高级应用技巧提升
- BIM项目团队实施引导
- 全员BIM设计推广
- BIM工作室团队培养
- Improve basic BIM application skills of all members
- Improve advanced BIM application skills
- Guide to implementation of BIM project team
- Promote BIM design to all members
- Develop BIM studio team training

2. 第二步：选择BIM培训对象
Step 2: Select BIM trainees

层级 Level	培训对象 Trainees
企业级BIM培训 Enterprise-level BIM training	行业和企业决策者、经营部负责人、中层管理干部、总建筑师、总工程师等 Industry and enterprise decision makers, principals of business departments, mid-level management staff, chief architects, and chief engineers
项目级BIM培训 Project-level BIM training	项目经理、设计主持人、工种负责人、专业设计师等 Project managers, design principals, specialty principals, and professional designers

企业级BIM培训目标
Enterprise-level training goals

- BIM技术对行业、企业未来发展价值宣讲
- BIM对项目竞标经营价值宣讲
- BIM对业务流程、质量控制、数据管理价值宣讲
- Promote BIM's value in the industry and enterprises' future development
- Promote BIM's value in project bidding and operation
- Promote BIM's value in business process, quality control, and data management

BIM培训人员
BIM trainees

- 行业、企业高层，决策者
- 企业经营部、中层管理干部
- 企业中层管理干部、主任设计师等业务骨干
- Top decision makers of industry and enterprises
- Mid-level management staff in business departments of enterprises
- Business backbones such as middle-management staff of enterprises, and chief designers

项目级BIM培训目标
Project-level BIM training goals

- 全员BIM基本应用技能普及
- BIM高级应用技巧提升
- BIM项目团队实施引导
- 全员BIM设计推广
- BIM工作室团队培养
- Popularize basic BIM application skills to all members
- Improve advanced BIM application skills
- Guide BIM project team implementation
- Promote BIM design to all members
- Develop BIM studio team training

BIM培训人员
BIM trainees

- 一线设计师
- 有应用基础的一线设计师
- 项目负责人、项目设计师
- 企业中高层领导、项目及专业负责人、一线设计师
- BIM工作室团队成员
- Frontline designers
- Frontline designers having application base
- Project principals and project designers
- Mid- and high-level leaders of enterprises, projects and specialty principals, frontline designers
- BIM studio team members

3. 第三步：选择BIM培训机构
Step 3: Select BIM training institutions

层级 Level	培训机构 Training Institution
企业级BIM培训 Enterprise – level BIM training	BIM技术软件厂商、知名BIM咨询顾问服务公司及BIM经验丰富的国际国内BIM设计企业等 BIM technical software manufacturers, well known BIM consulting and service companies, and international and local BIM design enterprises with rich BIM experience
项目级BIM培训 Project – level BIM training	BIM软件经销服务商、专职BIM培训机构、知名BIM咨询顾问服务公司或企业自身的BIM组织机构等 BIM software distributors and service providers, full-time BIM training institutions, well known BIM consulting and service companies or BIM enterprise organizations

企业级BIM培训目标
Enterprise-level training goals

- BIM技术对行业、企业未来发展价值宣讲
- BIM对项目竞标经营价值宣讲
- BIM对业务流程、质量控制、数据管理价值宣讲

- Promote BIM's value in the industry and enterprises' future development
- Promote BIM's value in project bidding and operation
- Promote BIM's value in business process, quality control, and data management

项目级BIM培训目标
Project-level BIM training targets

- 全员BIM基本应用技能普及
- BIM高级应用技巧提升
- BIM项目团队实施引导
- 全员BIM设计推广
- BIM工作室团队培养
- Popularize basic BIM application skills to all members
- Improve advanced BIM application skills
- Guide BIM project team implementation
- Promote BIM design to all members
- Develop BIM studio team training

- BIM厂商
- 国际、国内同行BIM先行者
- 知名BIM顾问咨询服务提供商
- BIM manufacturers
- Local and international BIM industry leaders
- Well known BIM consulting service providers

- BIM软件经销服务商
- 专职BIM培训机构
- 知名BIM顾问咨询服务提供商
- 企业内部BIM组织机构
- BIM software distributors and service providers
- Full-time BIM training institutions
- Well known BIM consulting service providers
- BIM enterprise organizations

4. 第四步：确定BIM培训教材
Step 4: Select BIM training materials

层级 Level	参考培训教材 Training Materials for Reference
企业级BIM培训 Enterprise-level BIM training	"中国建筑信息模型标准框架研究"、"BIM总论"、"BIM第二维度"等 *Research of Chinese Building Information Modeling Standard Framework (CBIMS), BIM,* and *The Second Dimension of BIM*
项目级BIM培训 Project-level BIM training	"Autodesk Revit Architecture 201x建筑设计全攻略"、"Autodesk Revit MEP 2012 应用宝典"、"Autodesk Revit 2012 族 达人速成"等 *Autodesk Revit Architecture 201x Architectural Design, Autodesk® Revit® MEP 2012, Autodesk Revit 2012 Family: Daren Crash*

5. 第五步：制定BIM培训计划
Step 5: Develop BIM training plan

层级 Level	培训计划主要内容 Main content of training plan
企业级BIM培训 Enterprise-level BIM training	1) BIM培训日程安排BIM training agenda 2) BIM培训内容大纲BIM training content outline 3) BIM培训主讲、助教Speakers and assistants for BIM training 4) BIM培训地点、设备Place and equipment for BIM training 5) BIM培训组织纪律BIM training organizational discipline
项目级BIM培训 Project-level BIM training	

6. 第六步：BIM培训过程管理
Step 6: BIM training process management

层级 Level	培训过程管理 Training Process Management
企业级BIM培训 Enterprise-level BIM training	签到、授课、意见反馈、互动讨论 Registration, teaching, feedback, exchange and discussion
项目级BIM培训 Project-level BIM training	培训签到、授课辅导、课上练习、课后作业、反馈意见、内容调整、集中答疑 Register for training, teaching and coaching, class exercises, homework, feedback, content adjustment, centralized Q&A

管理小秘籍：员工填写自愿参加培训表，设计室主任担保培训费，专人负责培训考勤，出勤率与结业成绩结合。凡毕业者由总部承担培训费，成绩不及格者由室主任承担培训费，成绩优异者总部另行奖励。

Management tips: Employees will voluntarily participate in training, and director of the design office will guarantee the training fee. Designate someone to check attendance, as attendance will be linked to the final grade. Those who pass the test will have their training fees paid by for the headquarters, and for those who fail the test, the director of the design studio will have to bear the training fee cost. The top employees will be rewarded separately by the headquarters.

签到 Registration → 授课辅导 Teaching and coaching → 课后作业与练习 Homework and practice → 培训意见反馈 Feedback on training → 培训内容方式调整 Adjustment of training content and method → 集中答疑 Centralized Q&A

7. 第七步：BIM培训考核认证
Step 7: BIM training assessment and certification

层级 Level	培训考核认证 Training Assessment and Certification
企业级BIM培训 Enterprise-level BIM training	1.内部认证证书（参见内部样本） 2.官方认证证书（需参加BIM厂商、行业协会等的官方认证培训，并通过考试）（参见官方样本） 1. Internal certificate (refer to internal sample) 2. Official certificate (required to attend official certification training held by BIM manufacturers and industry associations, and pass the test) (refer to official sample)
项目级BIM培训 Project-level BIM training	

纪律考核 Discipline assessment → 课堂笔记 Written examination → 课后作业 Homework → 综合成绩 Comprehensive results → 颁发证书 Issue certificates

官方认证证书样本
Sample of official certificate

内部认证证书样本
Sample of internal certificate

8. 第八步：BIM培训归纳总结
Step 8: BIM training summary and conclusion

层级 Level	培训总结主要内容 Main Content of Training Summary
企业级BIM培训 Enterprise-level BIM training	1.培训情况的总结 2.培训资料收集和共享（线下温习、自学） 3.问题与改进方案 4.下期目标、培训计划期望 1. Training summary 2. Collecting and sharing training materials (offline review, self-learning) 3. Problems and improvement program 4. Expectations on goals and training plan for the next stage
项目级BIM培训 Project-level BIM training	

培训情况总结报告 Training summary report → 培训资料收集共享 Collecting and sharing training materials → 问题与改进方案 Problems and improvement program → 下期目标计划预测 Expectations on goals and training plan for the next stage

4.6 BIM 技术设计协同
BIM Technological Design Collaboration

BIM多方设计协同
BIM Multi-Party Collaboration

BIM多方协同——各阶段中各方参与的BIM协调同工作模式和要点。

BIM multi-party collaboration
All parties take part in the BIM collaborative working mode and key points at each stage.

1. BIM设计阶段的协同要点
Key points for collaboration at BIM design stage

1）参与方及关注点
设计单位：关注完成设计任务的三要素以及三者之间的相互影响；通过协同可以提高企业核心竞争力。

1) Participants and focuses
Design units: Focus on the three elements necessary to complete a design task and the relationship between them: the enterprise's core competitiveness can be improved through collaboration.

前提条件要求 Pre-Condition		采用传统模式下的结果 Outcome in Traditional Mode	采用协同模式下的结果 Outcome in Collaborative Mode
时间 ↓ Time ↓	质量 ↑ Quality ↑	成本 ↑ Cost ↑	成本 ↓ Cost ↓
时间 ↓ Time ↓	成本 ↓ Cost ↓	质量 ↓ Quality ↓	质量 ↑ Quality ↑
质量 ↑ Quality ↑	成本 ↓ Cost ↓	时间 ↑ Time ↑	时间 ↓ Time ↓

投资单位：关注设计的进度，方案的合理性，和工程内容的移交。
数字化移交将使业主尽早得到工程信息：

Investment units: Focus on design schedule, program reasonability, and project handover:
Digital handover will enable owners to obtain project information as soon as possible:

数据移交：主要是在业主和各参建方之间的传递，工程信息的传递贯穿整个生命周期。

Data handover: Data is mainly transferred between owners and all participants, and the transfer of project information will run through the whole lifecycle.

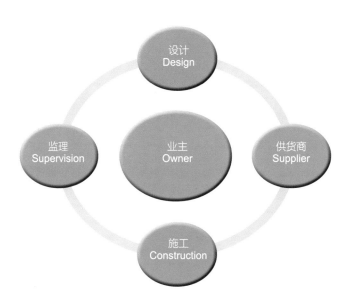

2）传统的协同模式和BIM协同模式的优缺点比较

2) Comparing the advantages and disadvantages between traditional and BIM collaborative modes

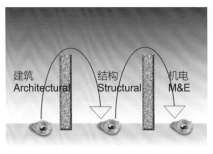

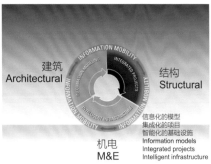

"传统协同模式，信息沟通犹如隔墙扔包裹"……

"In the traditional collaborative mode, information communication is like throwing parcels over walls"…

"BIM协同模式"实现信息的共享，及时沟通……

"BIM collaborative mode" enables information communication to be shared in a timely manner...

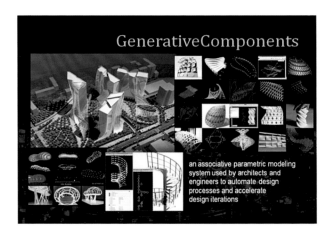

3）目前BIM协同存在的主要问题

行业缺乏统一标准，各软件厂商数据传递存在丢失，软件对计算机终端要求过高等。

3) Major limitations of current BIM collaboration

There is no uniform industrial standard, data loss occurs during data transfer among software manufacturers, and software is beyond the processing capacity of computers.

2. BIM多平台协同
BIM Multi-Platform Collaboration

BIM多平台协同——
BIM与各相关的设计平台之间的协同方式。

BIM multi-platform collaboration
Collaborative mode between BIM and each related design platform.

1）BIM与传统二维平台的协同要点
（1）传统二维设计平台的特点
信息的重复录入，信息关联性较差，后期信息调用较难，表达不直观。
（2）协同方式
建立同时管理BIM和传统二维文件的管理平台，建立衔接标准。
（3）运用范围
目前还不能全部脱离传统二维设计的企业和项目。
使用三维设计平台的企业与使用二维设计平台合作的项目。

2）BIM与参数化的协同要点

1) Key points for BIM collaboration with conventional 2D platforms
(1) Features of conventional 2D design platforms
Duplicated information input, poor information relevance, difficulties in information recall, expression not visualized.
(2) Collaboration mode
Establish a management platform to simultaneously manage BIM and conventional 2D documents, and set up a convergence standard.
(3) Applicable to
Enterprises and projects are not able to completely separate themselves from conventional 2D design.
Projects implemented by the enterprises adopting 3D design platforms and 2D design platforms.

2) Key points for BIM collaboration with parameters

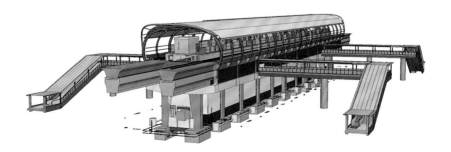

（1）参数化特点

程序建模、设计修改快捷方便。多方协作时，易统一标准。

（2）协同方式

采用参数化进行方案概念设计，BIM平台进行设计深化。

采用参数化进行方案表皮设计，BIM平台进行内部设计。

采用参数化进行工程后续设计，BIM平台进行深化设计。

（3）运用范围

异形建筑

大型复杂建筑

地铁、城铁、机场、车船码头等现代化的公共建筑

一般性建筑

3）BIM与绿色节能的协同要点

（1）绿色节能特点

绿色建筑是指在建筑的全寿命周期内，最大限度地节约资源（节能、节地、节水、节材）、保护环境和减少污染，为人们提供健康、适用和高效的使用空间，与自然和谐共生的建筑。

(1) Parametric features

Program modeling; enable rapid and easy design and modification. Easy to unify standards during collaboration.

(2) Collaborative method

Adopt parameters for concept design of plans, and BIM platforms for detailed design.

Adopt parameters for surface design of plans, and BIM platforms for internal design.

Adopt parameters for project's subsequent design, and BIM platforms for deepen design.

(3) Applicable to

Special-shaped buildings

Large and complex buildings

Modern public buildings, such as subways, urban transit rails, airports, and piers

Commom building

3) Key points for BIM collaboration with green energy-saving

(1) Features of green energy-saving

Green buildings are those that save resources (energy, land, water, and materials), protect the environment, and reduce pollution as much as possible during the whole lifecycle to provide people with healthy, applicable, and efficient space for use, and to be harmonious with nature.

（2）协同方式

BIM设计系统与各种分析、计算、模拟系统的集成和数据互用。

市政厅City Hall (Greater London Assembly)

(2) Collaborative mode

Integration and data exchange between the BIM design system and various analysis, calculation, and simulation systems.

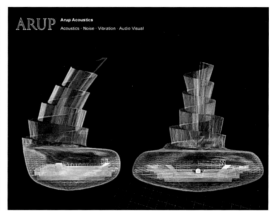

Foster and Partners/Arup

材质 / 照明 Materials / Lighting

University of Connecticut Information Technology Engineering Building
康涅狄格信息技术工程大厦
Burt Hill Kosar Rittelmann Associates

光线照度分析
Lighting Luminance Analysis

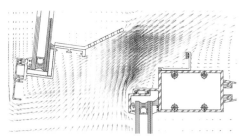
综合设计分析
Integrated Design Analysis

（3）运用范围
方案、初设和施工图设计阶段

(3) Applicable scope
Scheme design, preliminary design, and working design stages.

3. BIM多专业协同
BIM Multi-disciplinary Collaboration

BIM多专业协同——设计环节各专业之间以及专业内部的协同模式。

1）BIM工作模式的要点
（1）专业间协同设计模式
专业间通过文件链接将所需的专业设计汇总到一起。

BIM multi-disciplinary collaboration – Collaborative mode between specialties at the design stage.
1) Key points of BIM working mode
(1) Multi-disciplinary collaborative design mode
All specialties collect the required professional design by document links to work together.

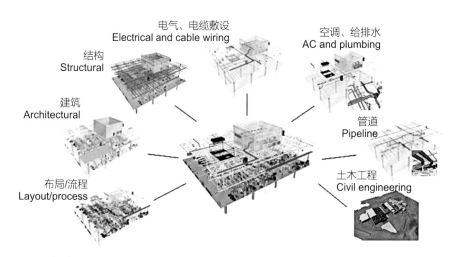

（2）专业内部协同设计模式

多人通过共享文档共同设计及编辑一个三维模型组合。

对于中大型项目，专业内需要先拆分，后通过文件链接汇总到一起。

(2) Inter-specialty collaborative design mode

People simultaneously and collaboratively design and edit a 3D model composite by sharing documents.

For medium and large projects, the disciplines should first be divided up and then collated together by document links.

矩阵式参考协同工作模式用于团队固定，专业固定的项目，长期、分撒项目
Matrix reference collaborative working mode, for projects with fixed teams and professionals as well as long-term and ad hoc projects

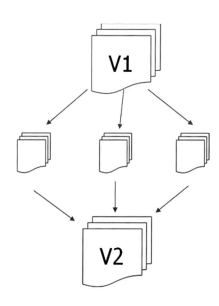

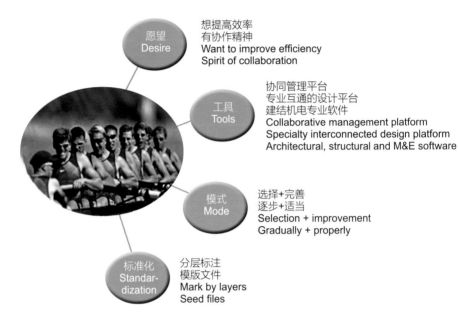

协同的要素　Collaboration elements

愿望 Desire: 想提高效率 / 有协作精神 / Want to improve efficiency / Spirit of collaboration

工具 Tools: 协同管理平台 / 专业互通的设计平台 / 建结机电专业软件 / Collaborative management platform / Specialty interconnected design platform / Architectural, structural and M&E software

模式 Mode: 选择+完善 / 逐步+适当 / Selection + improvement / Gradually + properly

标准化 Standardization: 分层标注 / 模版文件 / Mark by layers / Seed files

2）协同平台的技术实施方案
协同设计软件系统的搭建和配置

2) Technical implementation plans for collaborative platforms
Building and configuring a collaborative design software system

	作用 Function	能力 Capability	兼容 Compatibility
设计平台 Design platform	基于相同设计平台，完成各专业的设计内容 Complete design of each specialty based on the same platform	支持二维和三维设计，能够连接数据库 Support 2D and 3D design, and can be connected to the design database	兼容目前业界的各种图形、文字、图像格式 Compatible with all graphics, text, and image formats currently available in the industry
管理平台 Management platform	保证适当的人、在适当的时间、找到适当的信息/文件 Ensure the proper person can find the right information/documents at the right time	管理企业人员角色、访问权限及设计过程中所创建的所有文件 Manage roles of enterprise personnel, access rights, and all documents created during design	与项目管理、档案管理、ERP等衔接 Convergence of project management, archives management, and ERP

建立新型的多专业协同设计工作模式和作业流程

Establish new multi-discipline collaborative design work mode and flow

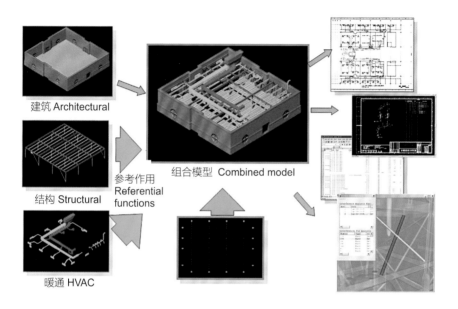

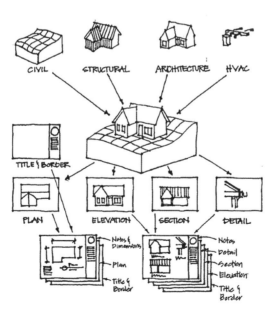

参考文献

1."绿色建筑评价标准（GB 50378）"；

2."中国建筑信息模型标准框架研究"；

3."BIM总论"；

4."BIM第二维度"；

5."Autodesk Revit Architecture 201x建筑设计全攻略"。

References

1. Evaluation Standard for Green Building (GB 50378);

2. Research of Chinese Building Information Modeling Standard Framework (CBIMS);

3. BIM general;

4. The Second Dimension of BIM;

5. Autodesk Revit Architecture 201x Architectural Design.

5

BIM技术在
设计项目方面的推广

BIM Promotion in
Design Projects

序号 No	步骤 Step	BIM设计项目方面的八个推广步骤 The eight extension steps of BIM design project
1	第一步（Step 1）	BIM项目目标 BIM project goals
2	第二步（Step 2）	BIM人员配置 BIM staffing
3	第三步（Step 3）	BIM硬件及软件配置 BIM hardware and software configuration
4	第四步（Step 4）	BIM计划和措施 BIM plan and measures
5	第五步（Step 5）	BIM技术培训 BIM technical training
6	第六步（Step 6）	BIM过程质量管理 BIM process quality management
7	第七步（Step 7）	方案、初设、施工图BIM技术应用点 BIM technological focus on schematic design, preliminary design and working drawing design
8	第八步（Step 8）	BIM成果总结 Summary of BIM outcome

5.1 BIM设计项目方面的八个推广步骤
Eight Steps of Promoting BIM in Design Projects

1. 第一步：BIM项目目标
Step 1: BIM project goals

目标类型 Type of goal	主要内容 Main content
技能目标 Skill target	掌握软件基本技能（多种BIM软件的数据交互）； Have command of basic software skills (multiple data exchange of BIM software);
人才目标 Talent target	培养BIM设计人才（BIM设计师、BIM经理）； Cultivate BIM design talent (BIM designers and BIM managers);
成果目标 Outcome target	实现BIM设计成果（积累资源，提高效率及质量，节约成本）。 Achieve BIM design results (accumulate resources, improve efficiency and quality, and save cost).

2. 第二步：BIM人员配置
Step 2: BIM staffing

根据项目阶段、项目规模、复杂程度，进行BIM团队人员配置。
Establish BIM team staffing based on project phase, scale and complexity.

<table>
<tr><td colspan="4" align="center">BIM团队人员构成
BIM Team Staffing</td></tr>
<tr><td>角色
Role</td><td>资格
Qualifications</td><td>职责
Responsibilities</td><td>基本配置人数
Staffing number</td></tr>
<tr><td>BIM经理
BIM manager</td><td>1.专业负责人及以上人员；
2.通过相关BIM软件资格认证；
3.全程参与过2个以上BIM平台的设计项目；
4.多专业综合组织能力较强。
1.Specialized director and above;
2.Has BIM software qualification certificate;
3.Has been involved in at least two BIM-based design project;
4.Strong multidisciplinary organization ability</td><td>1.统一整个团队使用BIM的意识；
2.BIM标准在项目中的落实；
3.协助项目负责人进行人员调配、分工及进度控制；
4.协调资源，解决高阶技术问题。
1.Ensure the entire team knows how to use BIM;
2.Implement BIM standards in projects;
3.Coordinate with project director in staff allocation, work division and schedule control;
4.Coordinate resources and solve high level technical issues</td><td>1</td></tr>
<tr><td>BIM设计人
BIM designer</td><td>1.熟练掌握BIM软件；
2.了解BIM流程和标准。
1.High command of BIM software
2.Understand BIM flow and standards</td><td>使用BIM软件进行相关设计工作。
Implement BIM software for design work</td><td>若干
Several</td></tr>
<tr><td>BIM协调员
BIM coordinator</td><td>1.具有专业设计背景或多年BIM项目实施经验；
2.精通多种BIM软件；
3.深度理解项目级BIM流程及标准。
1.Has professional design background or years of BIM project experience
2.Proficient in various BIM software
3.Profound understanding of project-level BIM process and standards</td><td>1.指导监督BIM标准执行；
2.负责提供项目所需库；
3.解决项目所有技术问题；
4.全权负责项目阶段归档及整理。
1.Instruct and supervise implementation of BIM standards
2.Provide component database for projects
3.Solve all project technical issues
4.Solely responsible for archiving and compiling project documents</td><td>2</td></tr>
</table>

133

（续表）

BIM团队人员构成 BIM Team Staffing			
角色 Role	资格 Qualifications	职责 Responsibilities	基本配置人数 Staffing number
BIM技术员（相当于传统制图员）BIM technical staff (equal to cartographer)	1.非常熟练的软件操作能力；2.对BIM新技术有较高敏感度。1. Highly skilled in software operation 2. Highly sensitive to new BIM technology	辅助BIM设计人完成BIM基础建模、尺寸标注、布图打印等工作。Assist BIM designers to complete BIM basic modeling, dimensioning, drawing, printing, etc.	可选 Optional

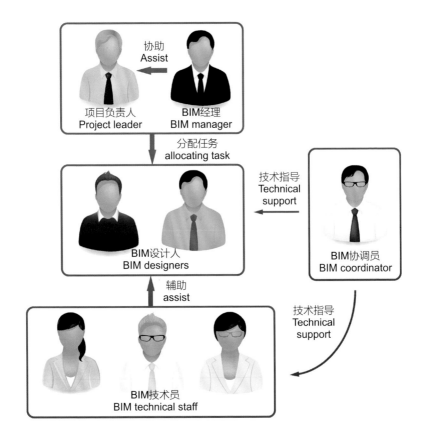

3. 第三步：BIM硬件及软件配置
Step 3: BIM hardware and software configuration

根据项目复杂程度及成本情况配置合理的软硬件设备
Rationally configure software and hardware devices based on project cost and complexity.

1）硬件配置

1) Hardware configuration

序号	主要配件	基本配置	推荐配置
1	CPU	Intel酷睿i5	Intel酷睿i7
2	内存	4Gx2	8Gx2
3	硬盘	普通硬盘	固态硬盘
4	显卡	独立显卡	独立显卡
5	显示器	1台19寸宽屏	2台19寸宽屏
	参考价	约6000元	约1万元

SN	Main components	Basic configuration	Recommended configuration
1	CPU	Intel Core i5	Intel Core i7
2	Memory	4Gx2	8Gx2
3	Hard disk	HDD	SSD
4	Graphics card	Discrete graphics	Discrete graphics
5	Display	1×19" widescreen	2×19" widescreens
	Reference price	Approx. RMB 6,000	Approx. RMB 10,000

2）软件配置（详见第6章内容）

常用BIM设计软件：Revit、Navisworks、Microstation、CATIA、MagiCAD、ArchiCAD等

2) Software configuration (refer to Chapter SIX)

Commonly used BIM design software: Revit, Navisworks, Microstation, CATIA, MagiCAD, ArchiCAD, etc.

4. 第四步：项目级BIM实施技术措施
Step 4: BIM plan and measures

1）按照项目情况制定进度计划
2）按照项目情况制定技术措施

1) Set schedule plan based on project situation
2) Work out technical measures based on project situation

序号 SN	技术措施 Technical Measures	主要内容 Major Content
1	项目标准 Project standard	构件库及项目样板 Component database and project templates
2	交付成果 Deliverables	最终交付格式 Final delivery format
3	模型拆分 Model split	模型进行二次拆分 Secondary split of model
4	搭建深度 Depth of modeling	根据成果要求确定模型深度 Confirm depth of modeling based on submission requirements
5	数据交互 Data exchanges	构建专业间数据交互模式 Build multidisciplinary data interaction mode
6	项目归档 Project archive	项目成果的审核归档 Review and archive project results

5. 第五步：BIM技术培训
Step 5: BIM technical training

根据项目类型和人员（已接受技术基础培训）配置，进行有针对性的BIM专项技术培训。专项技术培训的五个部分：

1）全专业协同设计培训
2）按项目类型进行建模技巧培训
3）常用布图、打印培训
4）详图深化设计培训
5）全专业碰撞检测培训

Provide targeted BIM special technical training based on the project and staffing (with basic technical training). Special technical training includes the following five parts:

1) Comprehensive training on professional collaborative design
2) Training on modeling skills by project type
3) Training on common drawing and printing
4) Training on comprehensive detail design
5) Comprehensive training on professional collision detection

6. 第六步：BIM过程质量管理
Step6: BIM process quality management

序号 SN	质量管理 Quality Management	控制点 Control point	责任人 Principals
1	项目手册的执行情况 Implementation of project manual	实时监控 Real time monitoring	BIM协调员 BIM coordinator
2	协同数据流 Collaborative data stream	阶段性审核 Phase review	BIM经理 BIM manager
3	交付成果 Deliverables	最终审核 Final review	BIM经理及项目负责人 BIM manager and project leader

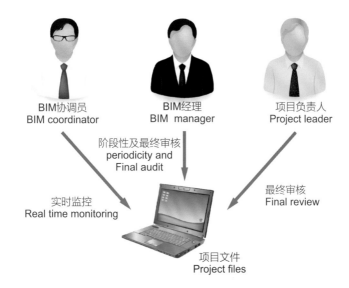

7. 第七步：方案、初设、施工图BIM技术应用点
Step7: BIM technological focus on schematic design, preliminary design and working drawing design

序号 SN	阶段 Phase	BIM技术应用点 BIM Technological Focus
1	方案阶段 Scheme design	经济技术指标控制，节能分析，多方案比选等 Control of economic and technological indexes, multi-scheme comparison
2	初设阶段 Preliminary design	全专业优化设计，消防疏散模拟，能耗分析，设计概算等 Full professional optimization design, fire evacuation simulation, energy consumption analysis, and design estimates
3	施工图阶段 Working drawing design	图纸生成，工程材料清单的自动统计，碰撞检测等 Drawing production, automatic statistics of quantities and bills of materials, and collision detection

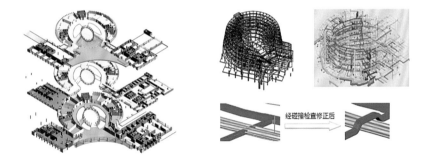

经碰撞检查修正后 Corrected by collision detection

8. 第八步：BIM项目总结
Step 8: Summary of BIM outcome

序号 SN	目标 Target	主要内容 Major Content
1	技能 Skill	软件兼容度、软件适用度、各平台间最优配置 Software compatibility, applicable software, and optimal configuration among all platforms
2	人才 Talent	BIM项目经理、设计师培养程度及数量 Qualification and quantity of BIM project managers and designers
3	成果 Outcome	设计质量与效率，资料库积累，成本情况 Design quality and efficiency, database accumulation, and cost
4	经验及困难 Experiences and difficulties	团队配合、软件技术、流程管控等 Teamwork, software technology, and process control and management

5.2 中国院2010年BIM应用部分项目
CAG's Implementation of BIM in 2010

1. 中国院 BIM 设计实践范围
CAG's practice scope of BIM

中国建筑设计研究院BIM实践范围：
- 选择不同类型的项目使用BIM技术，总结适合的方法与基本模版。
- 项目类型包括：住宅、办公建筑、酒店建筑、体育建筑、观演建筑、交通建筑、医疗建筑、规划设计、城市设计、室内设计、景观设计等。

CAG's Practice scope of BIM
- Select different types of projects using BIM technology, summing up proper methods and basic templates.
- Project types include: buildings for residential, office, hotel, sports, theater, transportation, medical, planning, and urban design, interior design, and landscape design.

2. 中国院 BIM设计案例实践线路
CAG's practice path of BIM design

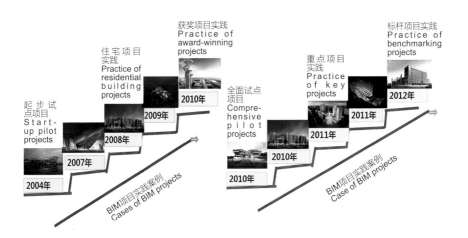

3. 中国院BIM设计应用部分项目
CAG projects using BIM design

某学院图书馆
An institute library

北京某公寓（住宅）
An apartment in Beijing
(residential building)

北京某会议中心
A conference center in Beijing

长白山某酒店
A hotel in Changbaishan

某服务中心
A service center

北京某办公楼
An office building in Beijing

5.3 中国院BIM应用代表案例——某金融商务中心
Typical Case CAG using BIM Technology: A Financial Business Centre

1. 某金融商务中心项目概况和获奖情况
Project Overview and Awards of a Financial Business Centre

1）项目BIM设计目标
（1）希望通过BIM技术，减少设计错误，提高设计质量；
（2）利用BIM三维优势，解决复杂体型，节省设计周期；
（3）通过信息传递，缩短后期建造时间，节约项目投资；
（4）计划培养熟练掌握BIM技术的设计人员；
（5）参加中国勘察设计协会BIM设计大赛；
（6）计划实现五个项目 BIM技术创新点。

1) Design goals of BIM project
(1) Use BIM technology to reduce design errors and improve design quality;
(2) Use BIM's 3D advantage to resolve complex body sizes and reduce design cycle;
(3) Shorten post-construction period and save project investment by information transmission;
(4) Plan to develop BIM designers to have strong command of BIM technology;
(5) Participate in BIM design competitions held by China Exploration & Design Association;
(6) Plan to achieve five BIM innovative highlights.

某金融商务中心项目BIM全专业设计应用

2）项目概况

项目名称：某金融商务中心。

建筑规模：约3万m²

建筑功能：城市商业中心

结构形式：地上为钢结构

设计模式：全专业BIM设计

Full-discipline design application of BIM in a Financial Business Centre

2) Project overview

Project name: A Financial Business Centre.
Construction area: approx. 30,000m²
Building function: Urban business center
Structure: Steel structure above ground
Design mode: Full-discipline BIM design

3）项目获奖情况

3) Awards

获中国勘察设计协会主办的第二届"创新杯"BIM设计大赛（2011）的四项大奖：
- 最佳BIM工程设计奖一等奖最佳BIM协同设计奖三等奖
- 最佳BIM应用企业奖最佳BIM建筑设计奖二等奖

Won four awards in the 2nd "Innovation Cup" for BIM Design (2011) sponsored by China Exploration & Design Association:
- First prize for Best BIM Engineering Design Award, third prize for Best BIM Collaborative Design Award
- Award for Best Application for BIM in Enterprises, second prize for Best BIM Building Design Award

2. 某金融商务中心五个BIM技术创新点
Five innovative highlights of BIM technology for a Financial Business Centre

1）创新点之一：建筑施工图100%采用BIM平台输出
1) Innovative highlight 1: Architectural working drawings fully adopt BIM platform for output

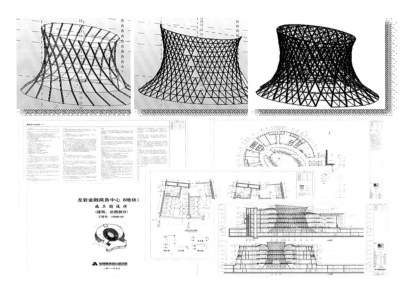

建筑专业使用Revit Architecture搭建复杂空间网架模型并实现100%BIM出图。
The architectural discipline uses Revit Architecture to build complex spatial grid models and fully adopts BIM for working drawing output.

2）创新点之二：Revit模型导入结构计算软件

2) Innovative highlight 2: Structural computation software is imported into to the Revit model

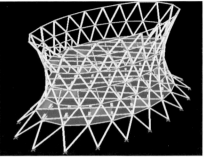

通过插件，实现将Revit模型单线图导入sap2000、PKPM。

Plug-ins are used to introduce single line drawings of Revit models into sap2000 and PKPM.

3）创新点之三：基于机械制造业技术的局部细化设计

3) Innovative highlight 3: Local detailed design based on machine manufacturing technology

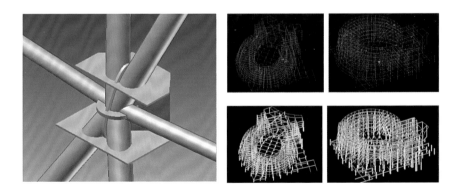

使用Inventor软件进行空间网架钢结构节点设计。

The Inventor software is used for node design of spatial grid structures.

4）创新点之四：模板图及配筋图的BIM绘图技术

4）Innovative highlight 4: BIM drawing technology in template and reinforcement drawings

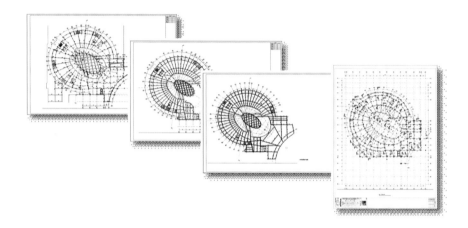

使用Revit Structure 绘制的模板图、配筋图。
Template and reinforcement drawings drawn by Revit Structure.

5）创新点之五：基于产品实际技术参数的设计技术

5）Innovative highlight 5: Design technology based on actual technical parameters of products

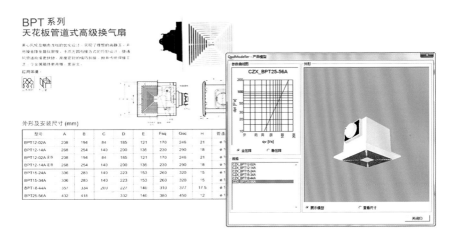

BPT系列

天花板管道式高级换气扇

离心风轮及蜗壳流线的优化设计，实现了理想的高静压，并将噪声降低至最低限度。卡爪方圆衔接方式的巧妙设计，使通风管道衔接更快捷，高度密封的精巧扣接，抛弃传统焊接工艺，令金属箱体更高雅、更安全。

外形及安装尺寸

机电专业使用机电BIM软件实现基于产品实际技术参数的设计，使得BIM模型与实际项目联系地更加紧密。

BPT

Ceiling duct ventilation fan

Optimized design of the centrifugal wind impellers to achieve ideal high hydrostatic pressure and minimize noise. Creative design of the claw square-round connection allows more rapid connection of air ducts. The highly sealed delicate fastening avoids using the conventional welding process so that the metal boxes can look more elegant and safer.

Profile and installation dimensions

The mechanical and electrical engineering designers use BIM software to achieve designs based on the actual technical parameters of products so that BIM models and actual projects can have closer links.

5.4 中国院 BIM应用经典案例——北京某信息产业基地
Classic Case of CAG using BIM Technology: Beijing XX Information Industry Base Project

1. 北京某信息产业基地项目概况及获奖情况
Project Overview and Awards of Beijing XX Information Industry Base Project

获中国勘察设计协会主办的第三届"创新杯"BIM设计大赛（2012）的三项大奖：最佳BIM协同设计一等奖、最佳BIM拓展应用奖、最佳BIM应用企业奖

项目名称：北京某信息产业基地项目

建筑规模：约20万m^2

建筑功能：综合办公

设计方式：全专业全过程BIM设计

Won three prizes in the 3rd "Innovation Cup" for BIM Design (2012) sponsored by China Exploration and Design Association. First prize for Best Collaborative BIM Design Award, Best Award for BIM Extension Application, Award for Best Application of BIM for Enterprises

Project Name: Beijing XX Information Industry Base Project

Construction area: approx. 200,000 m^2

Building function: Comprehensive office

Design method: Full-discipline and full-process of BIM design

2. 某国际信息港二期五个精细化设计
Five highlights of BIM-based refined design for A International Information Port (Phase II)

1) 精细化设计之一 ——多方案比选
1) Refined design highlight 1: multi-scheme comparison

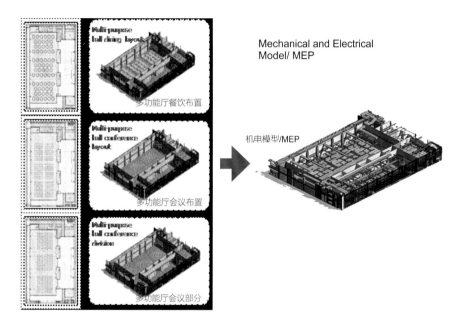

多方案比选
建筑先行，全专业共同参与，可视化协同设计减少专业间技术壁垒，协作也更为高效。

Multi-scheme Comparison
With construction going ahead, all the professionals worked together, using visual collaborative design to reduce technical barriers among enabling more efficient collaboration.

2）精细化设计之二——管线综合
2) Refined design highlight 2: Pipeline integration

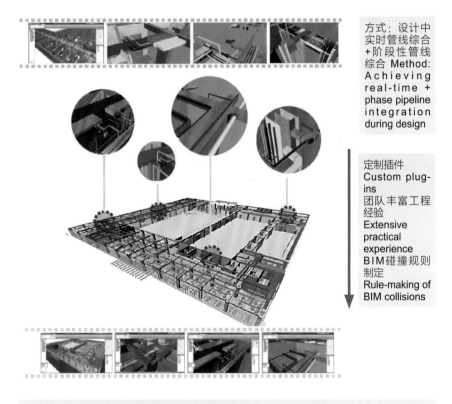

方式：设计中实时管线综合+阶段性管线综合 Method: Achieving real-time + phase pipeline integration during design

定制插件
Custom plug-ins
团队丰富工程经验
Extensive practical experience
BIM碰撞规则制定
Rule-making of BIM collisions

设计成果：主管零碰撞，支管碰撞数每楼层少于20个
Design outcome: No collisions of main pipes, and less than 20 branch pipe collisions on each storey

3) 精细化设计之三——三维的思考设计
3) Refined design highlight 3: Design with 3D methodology

三维可视化设计完美再现设计师设计意图，使业主更容易理解，方便沟通交流，实现有效决策。
3D visual design perfectly represents designers' ideas to facilitate owners' understanding, communication and exchange, and achieve effective decision-making.

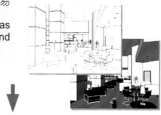

4) 精细化设计之四——绿色仿真模拟计算
4) Refined design highlight 4: Green simulation calculation

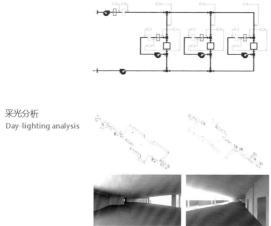

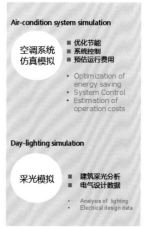

5）精细化设计之五——能耗及舒适度分析
5) Refined design highlight 5: Energy consumption and comfort analysis

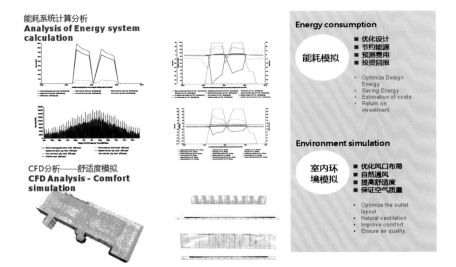

BIM技术设计应用软件汇总

Summary of BIM Technology in the Design of Application Software

6.1 BIM技术软件汇总之一(美国Autodesk公司)
Summary of BIM Tool Applications I (U.S.A Autodesk Inc.)

序号 NO	软件商 Vendor	软件LOGO Software Logo	软件名称 Software Brand	软件的主要功能 Major Function Description
1	欧特克 Autodesk		Revit	建筑、结构和机电专业的集成设计软件,适用于参数化建模、冲突检测、出图、报表生成,数据库支持、云端分析,支持多团队多专业的协同设计,API应用开发。 Integration of architectural, structural, and MEP professional design software, for parametric modeling, clash detection, drawing, report generation, database support, cloud analysis, supportscollaborative design, and API application development.
2			Navisworks Manage	设计到施工全过程的分析模拟软件,支持不同格式模型整合,碰撞检查、漫游、施工进度模拟与造价分析,API应用开发和数据移动客户端。 Whole process simulation software, supports different format model integration, clash detection, roaming, scheduling and cost estimation, API application development and mobile device.
3			Ecotect	建筑性能分析软件,可进行热工分析、光气象分析、声分析、风环境分析等。 Building performance analysis software, supports thermal analysis, light meteorological analysis, acoustic analysis, wind environmentanalysis, etc.
4			Civil 3D	土木工程设计软件,用于交通运输、土地开发和水利项目等基础设施领域的设计与文档编制。 Civil engineering design software for design and documentation of infrastructure and transportation, land development, and water projects, etc.
5			AutoCAD® Plant 3D	流程工厂的设计软件,基于等级的设计和标准零件目录库,帮助设计师高效创建管道,设备和支撑结构的布置流程。 Plant design software, based on specification-driven design and standard components library to streamline the placement of piping, equipment, and support structures.
6			Robot Structural Analysis	结构计算分析软件,利用Revit的互操作性简化共享结构分析模型和分析结果,辅助钢结构和钢筋混凝土结构的设计。 Structural analysis software, uses Revit® software interoperability to share structure analysis model and results, simply the design of steel and reinforced concrete structure.
7			Fabrication CAMduct	管道设计加工软件,拥有众多参数库以及压力级驱动的管道组件,可高效制作并安装HVAC建筑系统。 Pipeline design and processing software, contains parameters library, as well as the pressure driven pipeline components, efficiently product and install HVAC building systems.

BIM技术设计应用软件汇总
Summary of BIM Technology in the Design of Application Software

（续表）

序号 NO	软件商 Vendor	软件LOGO Software Logo	软件名称 Software Brand	软件的主要功能 Major Function Description
8	欧特克 Autodesk		AutoCAD Structural Detailing	结构详图和施工装配图设计软件，用于绘制结构草图、钢结构和钢筋混凝土结构，以及装配施工图。Structural details and construction assembly drawing design software, for drawing structure sketch, steel and reinforced concrete structures, as well as assembly construction drawings.
9			QTO	工程量统计软件，整合多方设计信息，基于BIM模型进行工程量统计与造价分析。Quantity Take-off software, integrates various design information, to generate bill of quantity and cost estimation based on BIM models.
10			Autodesk BIM 360/ Simulate 360	云端BIM应用解决方案，含基于云端的文件管理、3D模型浏览、云端协同以及云计算能力。Cloud BIM application solutions, including cloud-based file management, 3D model browsing, cloud collaborative, and cloud computing power.
11			Vault Professional	数据管理与协同软件，用于BIM工作流程的文档管理、数据管理、权限管理、流程管理、版本与操作记录管理。Data management and collaboration software, for BIM workflow document management, data management, authority management, process management, version and operating records management.
12			BIM 360 Glue	云端模型整合平台，支持网页及移动客户端模型显示与漫游，实现碰撞检测、施工配合和跨地区项目协同等功能。Cloud-based model integration platform, support web and mobile device with model browsing and roaming capabilities, clash detection, construction simulation and interregional projects collaborative functions, etc.
13			Infr Work	城市规划概念设计软件，创建面向土木工程、交通运输和基础设施的三维模型，基于GIS进行总图、城市规划设计与成果展示，支持云端及移动客户端模型展示与故事场景。Infrastructure conceptual design software, for creating a 3D model for civil engineering, transportation and infrastructure, GIS-based general plan, urban planning and showcase, support cloud and mobile device with model display and story scenes.

电话：010-8565 8800 (总机)　网站：www.autodesk.com.cn

157

6.2 BIM技术软件汇总之二(美国Bentley公司)
Summary of BIM Tool Applications 2 (Bentley Systems U.S.A Ltd.)

序号 NO.	软件商 Vendor	软件LOGO Software Logo	软件名称 Software Name	软件的主要功能 Major Function Description
1	Bentley		Project-Wise	工程信息管理和协同平台,实现对工作内容、工作流程以及工作标准、工作环境、工程标准库的集中管理。采用分级授权管理机制,实现工作流程的自动控制。尤其适用于远程异地协作。 ProjectWise is a scalable collaboration platform for project information management for connecting people and information across the distributed enterprise, core capabilities enable users to manage, find and share content, workflow, standard, workspace, project data etc. Hierarchical authorization management systems to achieve workflow management automatically. Especially suitable for long distance collaboration in difference location。
2			AECOsim Building Designer	面向BIM的建筑、结构、建筑设备及建筑电气分专业三维设计模块(软件),其中建筑设备涵盖了暖通、给水排水及其他低压管道的设计功能。可完成各专业从模型创建、图纸输出、统计报表、碰撞检测、数据输出等工作。 This BIM modeling software application includes tools for architecture, structural, mechanical, and electrical system 3D design, and construction equipment covers HVAC, water supply and drainage and pipe design. AECOsim Building Design delivery the modeling creation, drawing output, report, clash detection, data output etc.
3			Micro-Station	3D(包含2D)设计平台软件,提供面向BIM全三维设计环境,具有大数据量的处理能力,同时具有良好的兼容性和开放性,全面兼容DWG、SKP、3DS等常用工程内容格式。提供照片般真实的渲染和动画功能。 MicroStation is the world's leading information modeling environment explicitly for BIM. MicroStation can be used either as a software application or as a technology platform. MicroStation provides immersive interaction with 3D models and 2D designs, information-rich 3D PDFs and 3D plots. Moreover, its ability to integrate a comprehensive breadth and depth of engineering geometry and data from an unmatched range of CAD software, such as DWG, SKP, 3DS.
4			Bentley Map	Bentley Map是一款功能完善的GIS软件,能够解决组织和企业中基础设施地图,规划,设计,建造和运营维护问题。Bentley Map能够收集和整合不同来源的CAD和GIS数据,进行浏览、编辑、空间分析,输出报表和制图以及GIS信息共享,从而为规划和决策提供支持。Bentley 是一个全三维的软件平台,使用已有的CAD/GIS 数据和图像、点云数据,能够快速创建三维城市模型。尤其适用于BIM建筑园区规划。 Bentley Map is a fully featured GIS that is 3D by nature. It is designed to address the unique and challenging needs of organizations that map, plan, design, build, and operate the world's infrastructure. It supports the creation, persistence, maintenance, analysis, and sharing 2D/3D geospatial information. It is also ideal for developing custom GIS application. Especially suitable for BIM campus planning.

（续表）

序号 NO.	软件商 Vendor	软件LOGO Software Logo	软件名称 Software Name	软件的主要功能 Major Function Description
5			Prosteel Professional	三维钢结构绘图和详图软件，支持MicroStation® 和AutoCAD，实现复杂结构的三维建模，节点连接，并自动生成后期图纸和材料表，查看和对比不同时期模型和文档的修改情况。可以与工程设计环节中的其他软件共享模型数据，包括结构分析和设计软件（STAAD® 和RAM™）以及工厂设计软件（AUTO-PLANT®）等。 Prosteel is a 3D modeling environment for structural steel and metal work supporting your construction and planning tasks, working on MicroStation® & AutoCAD, you get an intuitive and integrated multi-material modeler perfectly suited to layout complex structures, produce shop drawing, assemble all your connections and manage your bill of materials. Prosteel integrates with other structural software STAAD® & RAM™ and Plant design software AUTO-PLANT®.
6	Bentley		STAAD.Pro	结构分析模块（软件），可进行各种设施的钢结构、混凝土结构、木结构、铝结构和冷弯型钢结构设计。符合10CFR Part 50、10CFR21、ASME NQA-1-2000 标准的核工业认证、时程弹塑性分析和缆索（线性和非线性）分析。STAAD.Pro 还可与STAAD.foundation 和 ProSteel 等其他 Bentley 产品相集成，实现结构设计阶段的全覆盖。 STAAD.Pro allows structural engineers to analyze and design virtually any type of structure through flexible modeling environment, advanced features and fluent data collaboration. The design features are included such as nuclear certification for 10CFR Part 50、10CFR21、ASME NQA-1-2000, time history and push over analysis and cable(Linear & non-linear) analysis. STAAD. Pro integrates with Bentley products such as STAAD.Foundation & ProSteel. STAAD.Pro is provided for integration with third party programs.
7			Bentley Navigator	三维设计校审模块（软件），可以实现信息模型的浏览、渲染、动画、三维校审、批注、碰撞检测、施工进度模拟和动态吊/安装模拟等功能，与ProjectWise相结合，可以实现设计、校审、批注、反馈工作流程自动控制。支持Ipad移动终端。 Bentley Navigator is used to review and analyze project information. With highly versatile 3D viewing capabilities, Navigator delivers a more intuitive i -Models review, render, animation, clash detection, project schedule simulation and dynamic animation etc, cooperation with Projectwise, Navigator can also automatically implement design, 3D proofreading, marking project operation workflow. Supporting iPad apps.

电话：+86 10 5929 7000　　邮箱：china.feedback@bentley.com

6.3 BIM技术软件汇总之三（芬兰Progman公司）
Summary of BIM Tool Applications 3 (Finland Progman Co., Ltd.)

序号 NO.	软件商 Vendor	软件LOGO Software Logo	软件名称 Software Name	软件的主要功能 Major Function Description
1	Progman	MagiCAD	MagiCAD Ventilation	通风系统设计软件，支持AutoCAD和Revit双平台，支持标准IFC格式，含完善的产品数据库，可进行管径自动选择，管道水力计算，噪声等级计算，碰撞检测，材料统计，标准平面图纸生成，自动生成并更新剖面图，设备工作状态点的可视化巡检。 Ventilation system design module; Available for AutoCAD and Revit platforms, IFC format compatible; containing extensive product databases; User definable Calculation functions for sizing, balancing and sound levels; Collision control; Automatic Bill of Material; Producing 2D plan drawings in accordance with local standard, create and update sections easily and automatically; System simulation and visualization of system working condition (display device preset values/damper positions).
2		MagiCAD	MagiCAD Heating & Piping	采暖和给水排水系统设计软件，支持AutoCAD和Revit双平台，支持标准IFC格式，含完善的产品数据库，可进行管径自动选择，散热器自动选型，管道水力计算，碰撞检测，材料统计；标准平面图纸生成，自动生成并更新剖面图，设备工作状态点的可视化巡检。 Heating/Cooling and piping system design module; Available for AutoCAD and Revit platforms, IFC format compatible; containing extensive product databases; User definable calculation functions for pipe & radiator sizing and balancing; Collision control, Automatic Bill of Material; Producing 2D plan drawings in accordance with local standard, create and update sections easily and automatically; System simulation and visualization of system working condition (display device preset values and valve positions/Kv values).

（续表）

序号 NO.	软件商 Vendor	软件LOGO Software Logo	软件名称 Software Name	软件的主要功能 Major Function Description
3	Progman	MagiCAD	MagiCAD Electrical	电气系统设计软件，支持AutoCAD和Revit双平台，支持标准IFC格式，含完善的产品数据库，设备自动查找/替换功能，用户自定义产品和二维图标批量入库；可进行碰撞检测，材料统计，短路计算；标准平面图纸生成，自动生成并更新剖面图，快速生成配电盘原理图，并且可以和平面图进行双向更新，具有与照明计算软件的接口（Dialux插件）。Electrical system design module; Available for AutoCAD and Revit platforms, IFC format compatible; containing extensive product databases, automatically device find and replace; Converting 2D symbols and 3D objects to MagiCAD symbols and objects; Collision control; Automatic Bill of Material and Short-circuit calculation; Producing 2D plan drawings in accordance with local standard, create and update sections easily and automatically; Generating schematic drawing; synchronizing plan drawing and schematic drawing, Connection to lighting calculation software (Dialux).
4		MagiCAD	MagiCAD Room	智能建模软件，快速搭建围护结构模型，自动预留孔洞，全面支持标准IFC格式，可与照度分析和能耗分析软件进行配合。Software for architecture design; Building architecture and structure model; Generating provision of voids automatically; Support with Standard IFC format; Cooperation with illumination analysis tools and energy analysis tools.
5		MagiCAD	MagiCAD Supporter and Hanger System	管道支吊架设计软件，包含多种综合支吊架形式；自动剖切/分析MagiCAD建立的BIM模型，自动布置；沿管线拷贝，查找/替换等编辑功能；自动生成二维平面图和大样图；可进行碰撞检测，统计材料；计算管道荷载，并自动生成结构计算书，利用计算结果优化支吊架选材，节约成本。Supporter and hanger System Design module; Containing mostly frequently used supporter and hanger types; Automatically analyzing MagiCAD BIM model, auto-installing functions; Editing Functions of copy along duct/pipes, find & replace; Producing 2D plan drawings and detailing drawings; Collision control and Automatic Bill of Material; Duct/pipe stress calculation and generating stress calculation report in accordance with structural mechanics; Using calculation result to optimize material (steel type) selecting, to achieve the goal of cost-saving.

电话：+86 10 877 30700　　邮箱：info.china@progman.fi

6.4 BIM技术软件汇总之四（北京理正软件公司）
Summary of BIM Tool Applications 4 (Beijing Leading Software Co., Ltd.)

序号 No.	软件商 Vendor	软件LOGO Software Logo	软件名称 Software Name	软件的主要功能 Major Function Description
1	北京理正软件股份有限公司 Beijing Leading Software Co., Ltd	理正软件	理正协同工作平台 Leading Collaborative Work Platform	二、三维协同管理软件，可管理各专业形成的图纸、文件和BIM数据，完成设计项目提资、会签、校审、成果提交、出图打印、归档等工作。 Collaborative Work Platform supports 2D and 3D co-design,and manages all kinds of drawing papers, files and BIM data of various disciplines, completes the process of cross-discipline data interaction, countersigning, review, printing and archiving, etc.
2		理正软件	理正协同CAD系列软件 Leading collaborative CAD Series Software	建筑、给水排水、暖通、电气设计、结构分析计算系列软件，可单独完成各专业的施工图绘制、复杂建筑结构计算分析、三维展示，并形成各专业BIM数据。 Leading collaborative CAD Series Software can independently complete the construction drawing, analyze complex Building Structure, display 3D model , and generate the BIM data.
3		理正软件	理正BIM建筑软件（Revit版） Leading BIM for Architecture (Revit version)	基于Revit软件开发的快速建模工具集软件，提供快速批量轴线、墙、门窗、台阶、车道、辅助族库管理、增强过滤、构件检索等辅助设计工具。 Based on Revit , It provides tools to build BIM model , such as create axis, walls, doors and windows, steps, ramps, family files management, enhanced filtering components, searching components,etc.
4		理正软件	理正Revit管线转图综合工具集 Leading toolset from CAD To Revit for pipelines	将机电CAD图纸通过该软件完成从dwg到Revit模型的转换，完成拾取生管线、管线变高、管线打断、管线自动生成连接件、多管标注等。 Using the software, electromechanical CAD drawings can be exported from dwg to Revit Model.It also provides following functions such as creating pipelines by selected lines, changing pipeline's path, breaking pipelines, automatic generation of connectors, labling multiple pipelines,etc.

（续表）

序号 No.	软件商 Vendor	软件LOGO Software Logo	软件名称 Software Name	软件的主要功能 Major Function Description
5	北京理正软件股份有限公司 Beijing Leading Software Co., Ltd	理正软件	理正总平规划及单体设计报批平台 Leading Platform for planning of general layout & single-building design.	实现控规规划设计、总平规划设计，在形成规划图纸的基础上，同时形成标准化数据，并通过数据库进行管理，可被城市规划GIS系统直接调用。 Support regulatory control planning and design, planning and design of general layout, on the basis of the formation of the plan drawings, and standardized data and database management, Called by urban planning GIS system.
6		理正软件	理正勘察设计系列软件 Leading exploration design software	通过贯穿野外数据采集、土工试验、勘察数据整理的全过程，实现勘察数据全生命周期管理，利用三维技术将勘察数据实现可视化展示。 Through the process of field data collection、soil testing and geotechnical data processing, it realizes the entire-lifecycle management of geotechnical data, and the visual exposition of geotechnical data by using the 3D technique.
7		理正软件	理正Revit施工工程资料管理软件 Leading Revit construction engineering data management software	基于Revit完成构件模型与施工工程资料之间的快速关联填写，构造构件与资料之间的实物与业务的关联性。完成施工过程中质量安全、工程监理、竣工验收等各类控制表单的填写和管理。 Build the association between model and engineering data based on Revit, and create a relationship between material object and technological process. Complete filling and managing of various types of control form in construction, such as quality, safety, engineering supervision, completion, acceptance
8		理正软件	理正P-BIM施工集成软件 Leading integrated construction software based on the P-BIM	基于BIM标准的施工集成软件，兼容IFC标准。完成基于IFC数据模型的进度计划编制、查询及调整；工程概预算编制；施工工程资料的填写、与构件的挂接、查询及追溯。三维操作模式，可分项单独应用，也可集成协同工作。 The integrated construction software based on the BIM standard, which is compatible with IFC standard. Complete building, inquiring and adjusting schedule based on IFC model, project budget preparation, filling construction engineering data, linking with model. 3D operation model can be divided into items alone and be integrated to collaboration.

电话：（+86）010-68002688　　邮箱：lzcehua@126.com

6.5 BIM技术软件汇总之五（北京互联立方技术服务有限公司）
Summary of BIM Tool Applications 5 (Beijing isBIM Technical Services Co., Ltd.)

序号 No.	软件商 Vendor	软件LOGO Software Logo	软件名称 Software Name	软件的主要功能 Major Function Description
1	北京互联立方技术服务有限公司 Beijing isBIM Technical Services Co., Ltd.	OnePoint	BIM协同工作平台 OnePoint	实现项目数据文档的集中存储、文档版本控制、文档完成状态（提资料、设校审、会签、出图）、跟踪团队讨论和文档修改，完整的日志记录等管理功能。 BIM Project Management and Collaborative Platform. Key features Including: Project data files centralized storage; documents version control; Document status management, including data extraction, design review /countersign, drawing generation and so on; Team discussion and document revision tracking; Complete log records management.
2		isBIM	BIM构件库管理软件 BIM Model Elements Manager	BIM设计资源库的集中存储和管理软件，可实现构件预览、构件类别和属性的维护、族文件的上传和下载调用管理。 Software used for BIM model element library centralized storage and management. Major functions provided including: model elements preview, component category and attributes maintenance, Autodesk Revit Family files upload / download and access management.

BIM技术设计应用软件汇总
Summary of BIM Technology in the Design of Application Software

（续表）

序号 No.	软件商 Vendor	软件LOGO Software Logo	软件名称 Software Name	软件的主要功能 Major Function Description
3	北京互联立方技术服务有限公司 Beijing isBIM Technical Services Co., Ltd.	isBIM	isBIMQS工程量计算软件 isBIMQS	IsBIMQS在已有建筑BIM模型上，对建筑、设备等构件加入工程量算量编码，从而按计算规则计算出构件的工程量。 Associate classification code with each assembly item in created BIM model and generate Bill of Material or Take-offs Report based on business logic as well as take-offs and cost estimating rules.
4		isBIM	isBIM工具集软件isBIM Toolkit	提供基于Revit开发的一系列设计工具，可实现辅助Revit快速建模、碰撞检查、图纸生成等功能。 A set of Autodesk Revit plug-ins tools to aid Revit users rapid modeling, collision detection and drawing generatio.

电话：（+86）010-58931083　　电子信箱：shanglvsheng@isbim.com.cn

6.6 BIM技术软件汇总之六（北京鸿业同行科技公司）
Summary of BIM Tool Applications 6 (Beijing Hongye Tongxing Technology, Ltd.)

序号 No.	软件商 Vendor	软件 LOGO Software Logo	软件名称 Software Name	软件的主要功能 Major Function Description
1	北京鸿业同行科技有限公司 Beijing Hongye Tongxing Technology Co., Ltd	hy	HYArch for Revit 建筑软件	基于Revit的建筑专业软件，可实现快速建模、编辑与标注，门窗表、族库管理与协同设计功能。Revit-based architecture modeling software, which can realize rapid modeling, editing and annotation; providing family management and collaborative design with MEP model.
2		hy	HYMEP for Revit 给水排水软件	基于Revit的给水排水专业软件，实现从建模、计算到出图全流程的设计辅助。提供器具布置与连接，给水、排水、消防系统设计与计算校核、标注与统计、族库管理与协同设计功能。Revit-based water supply and drainage software. Providing equipment layout and connection, hydraulic calculation and piping adjustment, fire protection design and calculation verification, family management and collaborative capability.
3		hy	HYMEP for Revit 暖通空调设计软件	基于Revit的暖通空调专业软件，实现从建模、计算到出图全流程的设计辅助。提供设备布置与连接，风系统、水系统、采暖设计、系统计算与管道调整、负荷计算、标注与统计、族库管理与协同设计功能。Revit-based HVAC software. Providing tools for wind system, water system, heating design, calculation and piping adjustment, load calculations, family management and collaborative capability.
4		hy	HYMEP for Revit 电气设计软件	基于Revit的电气专业软件，实现从建模、计算到出图全流程的设计辅助。提供强弱电设备布置与连接，电气系统图、照度计算、负荷计算、标注与统计、族库管理与协同设计功能。Revit-based electrical software, Providing layout and connection for electrical equipment, illumination calculation and verification, load calculations, family management and collaborative capability.
5		hy	HYMEP for Revit 族库管理软件	提供基于服务器的产品族库管理、查询、布置工具。内含丰富的本地化设备及阀件族，可满足BIM设计、选型、计算、统计、生成施工图及提资需求，并可任意扩充。Providing server-based family database management, query and layout tool. Containing a wealth of localized equipment and valves families which can meet the demands of BIM design, type selection, calculations, statistics, construction plans generation and the provision of funding, and can be expanded at option.

BIM技术设计应用软件汇总
Summary of BIM Technology in the Design of Application Software

（续表）

序号 No.	软件商 Vendor	软件 LOGO Software Logo	软件名称 Software Name	软件的主要功能 Major Function Description
6			HYEP 鸿业全年负荷及能耗分析软件	基于EnergyPlus核心，可完成全年动态负荷分析及空调系统能耗分析，可输出全年逐时能耗分析报表、方案优化对比报表等。 EnergyPlus-based core, which can complete the annual dynamic load analysis and energy consumption analysis for air conditioning system, and can output the annual hourly energy consumption analysis reports, program optimization comparison reports, etc.
7	北京鸿业同行科技有限公司 Beijing Hongye Tongxing Technology Co., Ltd		HYRoad-Leader 路立得BIM道路设计软件	全三维的市政道路设计和公路设计软件，能够有效地辅助设计人员进行地形处理、平面设计、纵断面设计、横断面设计、边坡设计、交叉口设计、立交设计、三维漫游和交通模拟等工作。 Full 3D design of municipal roads and highway design software, which can effectively assist the designing staff conduct d terrain following, graphic design, longitudinal section design, cross-sectional design, slope design, intersection design, interchange design, three-dimensional roaming, traffic simulation, and other jobs.
8			HYFPS 鸿业总图设计软件	基于BIM理念的三维总图设计软件，提供地形建模和分析、土方计算和优化、道路设计、竖向设计、管线综合、统计报表等功能 BIM concept-based 3D master plan design software, providing terrain modeling and analysis, earthwork calculation and optimization, road design, vertical design, pipeline integration, statistical reports and other functions.
9			HYSUN 鸿业日照分析设计软件	提供建筑建模，单点、沿线、平立面、阴影分析、日照圆锥和传统的日影棒图等分析功能。提供日照预评估和日照容积率计算功能。 Providing architectural modeling, single point, linear, flat facade, shadow analysis, sunshine shadow cone and traditional bar graphs and other analytical functions.Providing pre-assessment of sunshine and sunshine plot ratio calculation.

电话：010-88312202 /88312238   电子信箱：BIM@hongye.com.cn

7

BIM技术应用遇到的问题和政府影响

Problems in BIM Technology Implementation and the Role of Government

7.1 BIM技术应用遇到的问题
Problems in BIM Technology Implementation

1. BIM文件的法律责任问题
Legal liability of BIM files

1）BIM文件在交付政府相关部门及业主时的合法性及有效性问题；

2）BIM文件在项目使用各方之间传递、使用时的知识产权和连带责任问题；

3）BIM模型与二维电子文件及蓝图之间版本及内容的一致性问题；

4）各阶段BIM数据的有效性确认与修改权限问题。

1) Legality and validity in delivering BIM documents to relevant government authorities and owners;

2) Intellectual property rights and joint liability in the transmission and implementation of BIM documents among project parties;

3) Consistency between the BIM model and two-dimensional electronic files and blueprints.

4) Validation and modification permission of BIM data at all stages.

2. BIM技术规范和标准问题
BIM Technical Specifications and Standards

目前BIM技术处于探索阶段，国家和地方相关的BIM规范及标准还处于前期调研、研究阶段，或正在制定中，相对滞后；由于BIM数据需要依赖软件平台进行生产与传递，所以BIM规范及标准与软件体系的融合程度是标准落实的关键。

Currently, BIM technology is in the exploratory stage, and so the relevant national and local BIM specifications and standards are lagging behind, and in the preliminary research stages or being developed. Since the software platform is necessary for the production and transmission of BIM data, the integration level of BIM specifications and standards, and software systems determines the implementation of standards.

3. BIM的应用和交付深度问题
BIM Application and Delivery Depth

1）不同阶段、不同用途的BIM设计深度标准不同；

2）BIM总体设计深度比二维设计有较大提高；

1) The BIM design depth standard depends on stages and purposes;

2) Compared with 2D design, BIM significantly improves the overall design depth;

3）BIM阶段设计深度比二维设计有较大提高；

4）BIM协同设计模式（含工作内容划分、工作阶段划分等）与二维协同设计模式有较大改变；

5）BIM设计模式的人员与工作岗位设置相对二维设计模式有较大改变。

3) Compared with 2D design, BIM greatly improves the stage design depth;

4) Compared with the 2D design model, the BIM collaborative design mode (including job content work division) makes considerable changes;

5) Compared with the 2D design model, the BIM collaborative design model significantly changes staffing and operating posts.

4. BIM技术软件的不完善问题
BIM Technology Software Imperfections

1）目前BIM软件以国外软件为主，本土化程度不够，影响推广应用；

2）软件间接口不完善，BIM数据打通程度不够；不同公司、不同用途BIM软件之间暂时不能实现数据信息的完美传递；

3）软件的应用对人员素质和硬件设备的要求较高；

4）涉及国家安全、保密信息的企业或项目，需要有中国自主知识产权的BIM软件作为基础平台。

1) The widespread use of foreign software and insufficient localization of BIM software affects its implementation and promotion;

2) The smooth transmission of data information between BIM software of different firms and with different purposes is currently not available due to imperfections in the software interface and insufficient links in BIM data;

3) The software application results in demanding higher qualified staff and hardware equipment;

4) For the firms or projects related to national security and confidential information, BIM software with Chinese proprietary intellectual property rights must serve as the foundation platform.

5. BIM技术文件建设档案馆存档问题
Archiving of BIM Technical Documents Construction Archives

现阶段各地城市建设档案馆的文件存储使用的是蓝图微缩技术，部分省市采用电子报批、电子存档。目前缺少BIM技术文件存档的标准，同时现有建设档案馆数据库的硬件及软件条件不足以满足BIM存档的要求。

Since archiving standards for BIM technical documents are not currently available at this stage, each urban construction archive library uses blueprint micro technology to store files, while some provinces and cities have applied for electronic approval and archiving. In addition, the hardware and software conditions of existing databases of construction archive libraries cannot satisfy BIM archiving requirements.

6. BIM技术各种成本较大问题
High Cost of BIM Technology

推动BIM技术的发展与应用将会带来行业的进步与发展，但同时在设备更新、软件更新、基础数据库搭建、人员培训、人力资源等方面的投入，比传统的二维设计有所增加。

The progress and development of the construction industry is ensured by promoting the development and implementation of BIM technology, but at the same time, the investment in updating equipment and software, establishing the foundation database, training staff and human resources is higher compared with traditional 2D design.

7.2 需要政府、协会在推进BIM技术过程中解决的问题
Problems to be Resolved in Promotion of BIM Technology by Government and Associations

1. 进行BIM技术的宣传推广
Promotion of BIM technology

1）建议举办BIM比赛、展览、交流研讨会；

2）在政府投资的项目，或有影响力的地标项目，倡导使用BIM技术。

1) Organize BIM competitions, exhibitions and exchange seminars;

2) Encourage the use of BIM technology in government-invested projects or influential landmark projects.

2. 制定BIM技术标准和技术措施
Formulate BIM technical standards and measures

1）建议成立"BIM技术专家委员会"；

2）出台初步的国家BIM技术标准和措施的指导性文件；

3）鼓励具有行业影响力和技术实力的企业，研究制定企业BIM标准，并推广应用。

1) Establish expert BIM technology committee;

2) Issue preliminary guidance documents of national BIM technological standards and measures;

3) Encourage influential firms with technical expertise to formulate and promote corporate BIM standards.

3. 制定BIM技术管理办法和奖励政策
Formulate BIM technical administration measures and incentive policies

1）对于使用BIM技术的项目，在报批、税费、资金上给予鼓励性政策；
2）建议编制"建筑工程BIM设计文件编制深度规定"；
3）建议成立"BIM技术仲裁委员会"。

1) Apply incentive policies on approval, tax and funding for projects using BIM technology;
2) Compile *Building Engineering BIM Design Documentation Depth Regulation*;
3) Establish BIM technology arbitration committee.

4. 出台"BIM技术报批制度"
Issue *BIM technology approval system*

建议分期、分阶段出台对于项目报批使用BIM技术的制度：
1）近期要求政府投资项目、地标项目等重大项目使用BIM技术进行电子报批；
2）中期要求达到一定规模（如公建项目2万m²以上及住宅10万m²以上）的项目使用BIM技术进行电子报批；
3）远期要求所有项目均使用BIM技术进行电子报批。

It is recommended to establish an approval system for projects to use BIM technology in stages:
1) Electronic approval should be used for major projects requiring government investment in the near future and landmark projects;
2) Electronic approval should be used for projects of a considerable scale (such as public projects at least 20,000 m² and residences at least 100,000m²) in the medium term;
3) Electronic approval should be used for all projects requiring BIM implementation in the long term.

5. 组织协调相关政府部门解决BIM审批及存档问题
Settle BIM approval and archiving by organizing and coordinating relevant government departments

1）制定BIM审批及存档相关政府部门（如规划、市政、档案等）审批协调的计划与步骤；
2）建议某些城市建设档案馆为试点，分区实现BIM竣工文件存档，形成新型智慧城市档案数据库；
3）在成熟之后，建议在全国逐步推广BIM文件审批及竣工文件存档。

1) Formulate approval and coordination plan and steps for BIM approval and archiving of related government authorities (such as planning, municipal and archives);
2) Choose some urban construction archives as pilot to achieve BIM document archiving and form a new intelligent urban archive database;
3) As it matures, the promotion of BIM document approval and archiving at the national level will be realized step by step.

6. 制定"BIM技术设计收费指导办法"
Formulate *Guidance Measures of BIM Technical Design Fees*

1）因采用BIM技术，设计质量提高及工作量增大，应制定相匹配的"BIM技术设计收费指导办法"；

2）建议制定BIM技术设计基本合同范本。

1) With the improvement of design quality and increased work load brought about by implementing BIM technology, a corresponding Guidance *Measures of BIM Technical Design Fees* should be formulated;

2) It is recommended to formulate a BIM technology design basic contract sample.

7.3 政府的BIM技术政策
Government BIM Technology Policies

1. 美国政府的BIM技术政策
The US government's BIM technology policies

美国政府的主要BIM应用单位，一般的做法是直接对提交的BIM模型及其关联的二维工程图纸进行审核，但在项目各个阶段，对BIM交付提出了不同要求，通常还要求同时提交与传统CAD交付类似的二维交付物，如平立剖图纸等。

The US government's general practice of major BIM application units is to approve the submitted BIM model and its related two-dimensional engineering drawings, but there are different requirements for BIM delivery at each stage, and usually there are demands for two-dimensional deliverables similar to traditional CAD deliverables, such as elevation and profile drawings.

2. 新加坡政府BIM技术政策
The Singapore government's BIM technology policies

1）从2013年6月1日起，新加坡规划部门开始实施，针对2万m²以上建筑方案必须采用BIM设计。

2）从2015年1月1日起，新加坡规划部门开始实施，针对0.5万m²以上建筑方案必须采用BIM设计。

3）新加坡政府计划投入约2.5亿新币，针对BIM设计部分实施相应补助。

4）针对节能建筑BIM设计，政府可适当放宽容积率控制政策。

1) The Singapore planning department stipulates that from Jun 1, 2013, BIM design must be incorporated into building programs over 20,000 m².

2) The Singapore planning department stipulates that from January 1, 2015, BIM design must be incorporated into building programs over 5,000 m².

3) The Singapore government plans to invest 2.5 billion Singapore dollars to subsidize BIM design.

4) The government will relax the floor area ratio control policy for BIM design in energy-saving buildings.

3. 中国政府BIM技术政策
The Chinese government's BIM technology policies

1）北京市城乡规划标准化办公室正在编制《BIM设计（民用建筑）基础标准》；

2）深圳市住房和建设局正在编制《深圳市工程设计行业BIM应用调研及发展指引》，深圳市发展局投资项目要求使用BIM技术；

3）2013年1月5日，上海市规划和国土资源管理局发布了沪规土资建【2013】8号《上海市建设工程三维审批规划管理试行意见》，从2013年3月1日起，对于如下项目推行建设工程设计方案三维审批规划管理：

（1）本市重点区域的公共设施项目（附区域详图）；

（2）景观风貌敏感区域24m以上且1万m²以上项目；

（3）优秀历史建筑、文物保护单位的立面改造项目；

（4）全市超高100m或10万m²以上项目；

（5）空间影响大的项目；

（6）市委市政府确定的重大工程或社会关注度高的项目等。

对于以上六类项目，在其核定规划条件、方案咨询、方案审批和竣工审批环节，建设单位和个人必须提交三维模型电子文件。

1) The Beijing Standardization Office of Urban-Rural Planning is compiling *BIM Design (Civil Buildings) Foundation Standard*;

2) The Shenzhen Housing and Construction Bureau is compiling *BIM Application Research and Development Guidance in Engineering Design Industry of Shenzhen*, and BIM technology must be used in projects invested by the Shenzhen Development Bureau;

3) On January 5, 2013, the Shanghai Municipal Administration of Planning, Land and Resources issued its investment construction No.8 document of 2013 called *Trial Opinions on Three-Dimensional Approval Planning Administration of Construction Projects of Shanghai*, which should come into effect on March 1, 2013 for the following projects;

(1) Public infrastructure projects in key areas of Shanghai (Regional detailed drawing attached);

(2) Projects with areas over 10,000m² and with landscape sensitive areas over 24m;

(3) Façade renovation projects of cultural relics protection organization and outstanding historical buildings;

(4) Projects with areas over 100,000m² or with elevations over 100m;

(5) Projects with considerable influence on space;

(6) Key projects confirmed by municipal governments or public-related projects.

For the above-mentioned projects, during the process of verifying planning conditions, consulting programs, approving programs and completing approval, the construction units and individuals should submit electronic documents of three-dimensional models.

7.4 BIM技术发展的思考
Reflections on BIM Technological Development

1. BIM技术发展的思考之一：适合设计企业发展的BIM应用格局
Reflection 1: BIM application pattern suitable for design firms' development

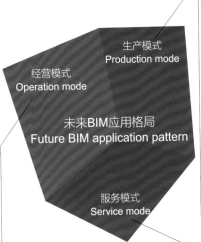

2）BIM的生产模式
2) BIM production mode

设计方式采用多样化生产模式（含传统的二维生产方式和现代的BIM三维生产方式），如采用三维生产方式需有配套的经济激励政策。It is designed to adopt diversified production modes (including traditional two-dimensional design and modern BIM three-dimensional design operation modes), such as 3D production method with matching economic incentive policies.

1）BIM的经营模式
1) BIM operation mode

设计合同采用多样化经营模式（含传统的二维设计和现代的BIM三维设计经营模式）供业主选择。
The design contract adopts diversified operation modes (including traditional two-dimensional design and modern BIM three-dimensional design operation models) for owners to choose.

3）BIM的服务模式
3) BIM service mode

通过BIM技术，提供高附加值服务模式，如：前期策划、方案比选、技术论证、施工配合、项目管理等服务模式。
Provide high value-added service modes using BIM technology, such as preliminary planning, scheme comparison, technological verification, construction coordination and project management.

175

2. BIM技术发展的思考之二：基于设计行业的BIM技术发展
Reflection 2: design industry-based BIM development

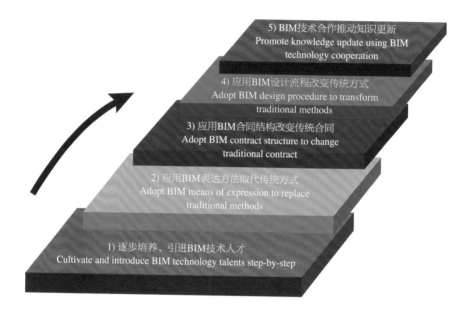

1）BIM人才：培养为主，引进为辅；

2）BIM方式：重视技术手段和表达方式的更新；

3）BIM合同：逐步从单一设计向设计总包、运营管理等合同结构的转变；

4）BIM流程：逐步创建新的BIM设计、校审、归档、输出流程；

5）BIM合作：开展与BIM技术相关各方的深度合作，推动知识更新。

1) BIM talents: mainly by cultivation, supported by introduction;

2) BIM method: Emphasize the updating of technological and means of expression;

3) BIM contract: Transform contract structure from single design to general design and operation management;

4) BIM procedure: Establish new procedures covering BIM design, review, archiving and output;

5) BIM cooperation: Implement deep cooperation among BIM technology-related parties and promotion knowledge update.

3. BIM技术发展的思考之三：BIM设计的六个阶段
Reflection3: six stages of BIM design

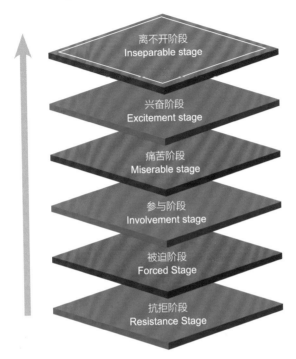

结论：
Conclusion:

通过以前的经验和目前BIM应用情况，我们认为：谁愈先采用BIM，谁愈早受益。正如1991年我们从手工绘图转变为计算机绘图一样，遇到了观念、习惯、效率、成本等各种困难，但是，只要认准方向，齐心协力，坚持不懈，就一定能提高企业的核心竞争力。

让我们以中国建筑设计集团的使命作为结束语："传承中华文化，打造中国设计；促进科技发展，引领行业进步。"让我们共同努力，积极推进 BIM 技术，创造美好的未来。

Judged from previous experiences and current BIM application, we believe that those who adopt BIM earlier will benefit earlier, just as in 1991, when we realized the transformation from manual drawing to CAD. At that time, we were confronted with obstacles from concepts, habits, efficiency and cost, but surely, with clear direction, concerted efforts and persistence, we can promote corporate core competiveness.

In conclusion here is the mission of CAG: inherit Chinese culture and establish Chinese design; push technological advancement and lead in industrial progress. We will work together to positively promote BIM technology and create a brighter future.

附录
Appendix

1. **中国建筑设计研究院**
   China Architecture Design & Research Group (CAG)

2. **美国Autodesk公司**
   U.S.A Autodesk Inc.

3. **美国Bentley公司**
   Bentley Systems U.S.A Ltd.

4. **芬兰Progman公司**
   Finland Progman Co., Ltd.

5. **北京理正软件公司**
   Beijing Leading Software Co., Ltd.

6. **北京互联立方技术服务有限公司**
   Beijing isBIM Technical Services Co., Ltd.

7. **北京鸿业同行科技公司**
   Beijing Hongye Tongxing Technology Co., Ltd.

中国建筑设计研究院（又称"中国建筑设计集团"）是国资委直属的科技型中央企业，是中国城镇建设领域中业务覆盖面广、科技创新能力强的建设科技综合集团，始终位居中国建筑设计企业前列。

中国建筑设计集团发源于1952年创建的中央直属设计公司，2000年由原建设部直属的建设部建筑设计院、中国建筑技术研究院、中国市政工程华北设计研究院和建设部城市建设研究院组建而成。

China Architecture Design &Research Group (CAG) is a large technology-intensive enterprise directly administered by State-owned Assets Supervision and Administration Commission of the State Council (SASAC). It is a leading engineering and consulting company in China with a wide range of businesses and a strong capacity for technology innovation in urban and rural construction field.

CAG evolved from the previous Central Design Company founded in 1952, and was constituted as its current form in 2000 by consolidating several large and influential companies - Architectural Design Institute of the former Ministry of Construction (MOC),

目前，中国建筑设计集团具有工程设计、城乡规划编制、文物保护工程、勘察设计、工程咨询、工程总承包等多项专业甲级资质；形成了城镇规划设计、民用建筑工程设计与咨询、基础设施与公共服务设施建设、建筑历史与文化遗产保护、节能环保、防灾减灾等技术领域的研究开发与技术服务的专业团队，成为中国建设科技领域的领跑者；在城镇规划、建筑设计、市政工程、标准规范、建设信息、工程咨询、室内装饰、风景园林等领域为中国乃至全世界的建筑及市政工程建设提供一体化的专业服务。

60年来，中国建筑设计集团先后设计完成了北京火车站、中国美术馆、国家图书馆、北京国际饭店、外交部办公楼、国家体育场（鸟巢）、首都博物馆、故宫保护、长城保护、引滦入津、西气东输、南水北调和长江三峡库区环境保护等标志性项目，"中国院"的品牌形象在国内外得到广泛认可。

China Building Technology Development Center, North China Municipal Engineering Design &Research Institute and China Urban Construction Design &Research Institute of MOC.

At present, CAG holds various A-level professional qualifications for engineering, urban and rural planning, cultural relic preservation, engineering survey and design, engineering consultation and EPC. CAG is respected as pioneer and leader in the sector for its strong R&D capacity and its outstanding professional teams in the field of urban planning and design, civil building engineering design and consultation, infrastructure development, historical and cultural heritage preservation, energy efficiency and environment protection, disaster prevention and reduction, etc. It aims to provide integrated services for urban planning, architectural design, municipal infrastructure planning, code &standards development, construction information, engineering consultation, interior decoration and landscape architecture.

Over the past six decades, CAG has designed a number of renowned projects including Beijing Railway Station, National Art Museum of China, National Library of China, Beijing International Hotel, Office Building of the Ministry of Foreign Affairs, National Stadium (Bird's Nest), Capital Museum, the Forbidden City Preservation, the Great Wall Preservation, Luan River to Tianjin Project, West to East Gas Transmission, South-to-North Water Diversion Project, and the Environmental Protection of Three Gorges Reservoir Region. The brand of "CAG" has been widely recognized in China and abroad.

电话：（+86）010-57700800
邮箱：cag@cadg.cn
网址：http://www.cadreg.com.cn

Tel：（+86）010-57700800
Email：cag@cadg.cn
Web：http://www.cadreg.com.cn

美国Autodesk公司
U.S.A Autodesk Inc.

欧特克有限公司("欧特克"或"Autodesk")是专业和个人领域三维设计、工程及娱乐软件的领导者,为制造业、工程建设行业、基础设施业以及传媒娱乐业提供卓越的数字化设计、工程与娱乐软件服务和解决方案。自1982年AutoCAD正式推向市场以来,欧特克已针对最广泛的应用领域研发出多种设计、工程和娱乐软件解决方案,帮助用户在设计转化为成品前体验自己的创意。《财富》排行榜名列前100位的公司普遍借助欧特克的软件解决方案进行设计、可视化和仿真分析,并对产品和

Autodesk, Inc., is a leader in professional and personal 3D design, engineering and entertainment software for the manufacturing, building and construction, and media and entertainment markets. Since its introduction of AutoCAD software in 1982, Autodesk continues to develop the broadest portfolio of state-of-the-art software to help customers experience their ideas digitally before they are built. Fortune 100 companies as well as the last 15 Academy Award winners for Best Visual Effects use Autodesk software tools to design, visualize and simulate their ideas to save time and money, enhance quality and foster innovation for competitive advantage.

Autodesk has 16 research and

项目在真实世界中的性能表现进行仿真分析,从而提高生产效率、有效地简化项目并实现利润最大化,把创意转变为竞争优势。

欧特克全球拥有16家研发中心,超过3000名研发人员。其中位于中国上海的欧特克中国研究院是欧特克全球最大的研发机构。欧特克每年的研发投入维持在全球总收入的20%的比例。对研发的巨大投入和不懈追求赋予了欧特克强大的创新能力,并通过卓越的思想、技术和解决方案将这种能力带给用户。

欧特克总部位于美国加利福尼亚州圣拉斐尔市,在全球111个国家和地区建立了分公司和办事处。在全球范围内,欧特克已经向超过一千万以上的正版用户提供优秀的、针对其行业特点的软件产品和技术咨询服务,同时向超过3400个全球软件开发商提供技术支持。

development centers, more than 3000 R & D staff. Autodesk China Research Institute in Shanghai is the largest Autodesk world's R & D institutions. Autodesk annual R&D investment maintains a ratio of 20% of the total global income. The huge investment in R&D and relentless pursuit to give Autodesk powerful innovation ability, and this ability gives the clients excellent ideas, technologies and solutions.

Autodesk is headquartered in San Rafael, California, USA, with branches and offices in 111 countries and regions around the world. Worldwide, Autodesk has provided outstanding software products and technology consulting services to over ten million genuine users for their industry characteristics, and provide technical support to more than 3,400 global software developer.

电话:010-8565 8800 (总机)
网址:www.autodesk.com.cn

美国Bentley软件公司
Bentley Systems U.S.A Ltd.

"Bentley公司始终致力于为设计、建造和运营全球基础设施的企业和专业人员提供创新的软件及服务,促进全球经济和环境的可持续发展,提高生活品质。"

Bentley公司是一家全球领先的企业,提供促进基础设施可持续发展的综合软件解决方案,包括用于设计和建模的 *MicroStation*、用于项目团队协作和工作共享的 *ProjectWise* 以及用于"项目和资产数据管理的*eB*"——所有解决方案均包含一系列"数据互用"的专业应用程序,并辅以全球专业服务。

Bentley公司于1984年在美国创建,在超过50个国家/地区拥有3000余名员工,年营收超过5亿美元。自2003年以来,公司在研发和收购方面已经投入逾10亿美元。

Bentley AECOsim——BIM用于建筑、结构、空调、水、暖、电气等多专业全信息建模、出图、和工程量的计算,其专业的均衡性和专业间无缝的集成性及支持超大体量模型的优异性能为建筑业树立了成功典范。Bentley ProjectWise协同管理平台,是项目的工程信息中心,既能负担内部协作又能完成外部协作,是BIM不可或缺的管理平台。

国内外成功案例:首都机场3号航站楼、水立方游泳馆、上海巨人网络集团总部、香港新机场、伦敦奥运场馆等。

"Bentley's mission is to provide innovative software and services for the enterprises and professionals who design, build and operate the world's infrastructure — sustaining the global economy and environment, for improved quality of life."

Bentley is the global leader dedicated to providing comprehensive software solutions for sustaining infrastructure. Its solutions encompass the *MicroStation* platform for infrastructure design and modeling, the *ProjectWise* platform for infrastructure project team collaboration and work sharing, and the *eB* platform for infrastructure asset operations – all supporting a broad portfolio of interoperable applications and complemented by worldwide professional services.

Founded in USA in 1984, Bentley has more than 3,000 colleagues in 50 countries, more than $500 million in annual revenues, and since 2003 has invested more than $1 billion in research, development, and acquisitions.

Bentley's AECOsim Building Designer is a multidisciplinary information modeling software for architectural, structural, air conditioning, mechanical and electrical systems' plotting and engineering calculation. It is the best practice in the Building industry with its seamless integration among multidiscipline and excellent performance on hyper-modeling. Bentley ProjectWise is the only project collaboration and AECO information management software, able to responsible for both of internal and external collaboration. It is an essential management platform for BIM.

Successful cases in China and worldwide include: Beijing International Airport Terminal 3, Water Cube – The Beijing National Aquatics Center, Giant Interactive Group Inc.

附录
Appendix

伦敦瑞士投资银行、伦敦议会大厦、伦敦维多利亚地铁、伦敦Crossrail交通枢纽、埃及开罗石塔商业区等。在中国Bentley BIM完成了大量的地铁和商业区项目。

Headquarter, Hong Kong International Airport, London Olympic Stadium, The Leadenhall Building, Palace of Westminster, Victoria Underground Station, London Crossrail, Stone Tower Business Park, etc. Many metro and commercial area projects in China have adopted Bentley BIM.

电话：+86 10 5929 7000
邮箱：china.feedback@bentley.com
网站：www.bentley.com/zh-CN

Tel: +86 10 5929 7000
Email: china.feedback@bentley.com
Web: www.bentley.com/zh-CN

芬兰Progman公司
Finland Progman Co., Ltd.

芬兰普罗格曼（Progman）有限公司于1983年成立，总部位于芬兰。在过去30年里，公司专注于建筑设备专业尤其是暖通专业的计算机软件应用技术。公司研发的MagiCAD 系列软件涵盖了暖通空调、建筑给水排水、建筑电气、建筑智能建模等各个专业部分；经历了30年的发展，MagiCAD成为整个北欧建筑设备设计领域内主导和领先的应用软件，占有绝对的市场优势；同时，Progman公司也逐步发展成为一个全球

Progman Oy was founded in 1983 in Finland. In the past 30 years, Progman has gained an outstanding knowledge and experience of developing software (MagiCAD as trademark) for professional HVAC and electrical design. MagiCAD provides solutions ranging from heating & cooling, water supply & drainage to ventilation and structure design. Over the years, MagiCAD has become the de facto standard among Nordic HVAC design agencies, and a clear market leader in BIM solutions for HVAC and electrical design in the Nordic countries. MagiCAD is marketed and supported through Progman sales companies and partners in

性的建筑设备软件（MEP）供应商，法国、荷兰、俄罗斯、土耳其、新加坡、马来西亚、中国香港等国家和地区均成为Progman公司市场的主要组成部分。

2008年年初，芬兰普罗格曼有限公司北京代表处正式成立；2008年5月，MagiCAD系列软件第一个本地化中文版正式推出，MagiCAD软件在随后的两年中快速的成为国内机电类BIM软件的主流产品，用户涵盖国外知名的工程设计咨询公司在国内的分支机构或者分公司、国内综合性大型设计单位、行业大型设计单位、大型工程总承包单位、施工单位、设备安装公司等，涉及公用建筑设计行业、电力，冶金、石化工业建筑设计行业。

Sweden, Norway, Denmark, Estonia, Lithuania, Latvia, the Benelux, Russia, China, Turkey and Poland.

Progman's Beijing office was established in the beginning of 2008. MagiCAD's first localized Chinese version was officially released in May 2008. In the following two years MagiCAD has quickly become the leading professional MEP solution in the Chinese BIM software market, and its users include domestic large integrated design institutes, industrial design institutes, general contracting companies, construction companies, MEP installation companies, and subsidiaries of international consulting companies and the involved industries include building design, electricity, metallurgy, and petrochemical industry.

电话：+86 10 877 30700
网站：http://www.magicad.com.cn
邮箱：info.china@progman.fi

Tel: +86 10 877 30700
Web: http://www.magicad.com.cn
Email: info.china@progman.fi

北京理正软件公司
BEIJING LEADING SOFTWARE CO., LTD.

北京理正软件股份有限公司成立于1995年7月，现有员工400余人，95%以上具有大学本科以上学历，是国家高新技术企业、双软认证企业、北京"中关村科技园区"入区企业，是建设行业内规模最大的软件开发企业之一，在上海、广州、青岛设有三个办事处。2002年6月成立的北京理正人信息技术有限公司，作为北京理正软件股份有限公司的控股子公司，是国内少数能以完全自主版权平台技术为数字水利、数字国土、数字建设、数字规划、数字城管、智慧城市、智慧社区等多个行业和领域信息化系统建设提供专业开发和品质型咨询顾问服务的软件公司之一。

北京理正是建设部"产业化软件研发示范基地"，以及多行业、多地区信息化工程的研发基地。通过了ISO9001国际质量认证和CMMI（软件能力成熟度模型）3级评估。北京理正还是中关村科技园区百家创新型试点企业、北京市守信企业、中关村"瞪羚计划"五星级企业、"中关村企业信用双百工程"首批入选企业，同时也是国家BIM标准研究唯一全部课题总承担单位（6个课题、33个子课题）。

公司自成立以来始终专注于勘察设计行业信息化以及政府信息化领域的软件研发和信息化咨询服务，目前已形成涵盖企业管理信息系统、协同设计、CAD计算机辅助设计、政府信息化四大

Beijing Leading Software Co. Ltd. (hereinafter referred to as BLS) is a share holding company established in July 1995, which is a high-tech enterprise. It has over 400 employees of whom 95% with a bachelor degree (30% are master and doctor degrees). And BLS has three branches in Shanghai, Guangzhou and Qingdao. As the fully-owned subsidiary of Beijing leading Software Founded in June 2002 in Beijing, Beijing Leading people Information Technology Co., Ltd. is one of the few software enterprises owning completely copyright platform offers professional development and quality-based consulting services for digitation water conservancy, land, construction, city planning, city management, Intelligent city, Intelligent Community.

BLS is the 'Industrialization demonstration base' of information products in construction field of Ministry of Construction, it is the R&D basement for multiple industries and regions. It is also the innovative pilot enterprise in Zhongguancun Science & Tech Park, awarding 'Beijing trustworthy enterprise', 5-star enterprise in 'Zhongguancun Gazelle plan', the first batch of 'Zhongguancun enterprise credit trustworthy' selected enterprise. Passing ISO9001 and CMMI level-3 assessment, it is the only chief-unit responsible for the entire subjects of International BIM standard research, including six topics and thirty-three sub-topics.

Since founded in 1995, BLS focuses on both the exploration & design industry information and the software R&D, information tech consulting services in informatization of government. It has formed the comprehensive product system, containing enterprise management information systems, collaborative designs, CAD, engineering constructions process

附录 Appendix

为中国BIM落地提供全方位信息化解决方案

领域的全面产品体系，业务涉及建筑、规划、铁路、公路、市政、水利、电力、冶金、地矿、邮电、石油等众多行业，市场占有率一直名列前茅。

"北京理正"是勘察设计行业的著名品牌。企业还承担了多项国家科技攻关课题项目的研究，参编多个国家、地方和行业标准，有多个产品被列入国家火炬计划项目，连续多年获得建设部科技示范工程、北京市科技进步奖。

management and informatization of government. Business involving engineering, city planning, highway, municipal, water conservancy, electric power, metallurgy, mining, telecommunications, petroleum, construction and government industries, Market share has been among the best.

BLS is a famous brand in exploration & design industry. It undertakes a number of national science and tech research subjects, and co-authors national, local, industrial standards. Many products are involved in the National Torch Program, and it also obtains the Ministry of Construction Science and technology demonstration project and the Beijing Science and Tech Progress Award for many years.

电话：（+86）010-68002688
邮箱：lzcehua@126.com
网址：http://www.lizheng.com.cn

Tel：（+86）010-68002688
Email：lzcehua@126.com
Web：http://www.lizheng.com.cn

北京互联立方技术服务有限公司
Beijing isBIM Technical Services Ltd.

北京互联立方技术服务有限公司（以下简称isBIM）系香港盖德软件科技集团有限公司投资的BIM服务企业，与同属集团的北京东经天元软件科技有限公司（Autodesk中国工程建设行业总代理商）、北京北纬华元软件科技有限公司（Autodesk中国最大的软件销售服务商）并列为集团三大核心企业。

isBIM自创建伊始，就秉承与中国工程建设行业客户共同实现技术创新的宗旨，先后为中国建筑设计研究院、北京市建筑设计研究院、ZahaHadid Architects、中国建筑工程总公司各工程局、北京万通股份有限公司项目等

Beijing isBIM Technical Services Limited (hereinafter referred to as isBIM), a wholly owned subsidiary in BIM services of the Hong Kong KOI TAK Software Technology Group Co., Ltd., and belong to the same group of REL (Royal East Longitude, Autodesk China general agent for AEC industry), RNL (Royal North Latitude, the largest Autodesk software sales and service provider in China) tied for the three core businesses of the group.

isBIM since its inception, adhering to the purpose of working together with the Chinese construction industry customers to achieve technological innovation, has provided China Architecture Design & Research Group, Beijing Institute of Architectural Design, Zaha Hadid Architects, just name a few and much more other customers with BIM software

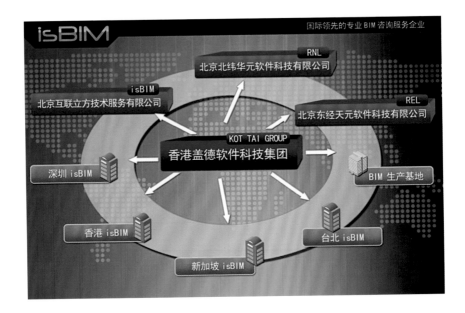

客户提供BIM软件选型咨询；建模；培训；项目引导；基于BIM的辅助设计、施工；基于BIM的ERP集成顾问；BIM企业标准建设等服务。

为了更好、更全面吸收国际先进的BIM应用模式与经验，集团于2009年、2012年分别投资组建了香港isBIM与新加坡isBIM，由香港isBIM主导的香港房屋署5D-BIM项目荣获英国皇家测量师学会2013年度创新大奖，开创了大中华区域将BIM技术应用应用于真实工程项目成本管理的先例。

为体现isBIM对行业技术发展承担的责任，isBIM参与编写的清华大学CBIMS系列第二册《设计企业BIM实施指南》已于2013年4月出版。为了BIM在行业中可持续发展，isBIM自2011年起，投资组建BIM生产基地，前瞻性的为未来行业BIM应用分工合作积极铺垫，力求从各个角度与行业共同迎接BIM引发的产业革命。

selectionconsulting , modeling, training, project guiding, aided design and construction based on BIM, BIM-based ERP integration consulting services, setting up BIM enterprise standards and other BIM value-added services.

In order to better and more comprehensive absorbing international advanced experience of BIM application mode, the Group invested in Hong Kong isBIM and Singapore isBIM in 2009, 2012 respectively. led by Hong Kong isBIM, Hong Kong Housing Department 5D-BIM project won the United Kingdom Royal Institute of Chartered Surveyors (RICS) 2013 innovation award, created a precedent of BIM technology was applied to real engineering project cost management in the Greater China region.

To embody isBIM 's responsibility for technical development of the industry, isBIM involved in the preparation of the Part II of the Tsinghua University CBIMS Series "Design Enterprises BIM Implementation Standards Guide" was published in April 2013. In order for BIM to sustainable development in the industry, isBIM since 2011, invested and established BIM production base, forward-looking actively paving the way for the industry BIM application to work in cooperation with a due division of labor in the future, and strive together with industry from different perspectives to meet the industrial revolution triggered by BIM.

电话：（+86）010-58931083
邮箱：shanglvsheng@isbim.com.cn
网址：http://www.isbim.com.cn

Tel：（+86）010-58931083
Email：shanglvsheng@isbim.com.cn
Web：http://www.isbim.com.cn

北京鸿业同行科技公司
Beijing Hongye Tongxing Technology Co., Ltd.

鸿业科技成立于1992年，是国内最早从事工程CAD设计软件开发的公司之一，致力于将建筑类专业与计算机软件技术相融合，为广大的设计人员提供工程CAD设计软件和城市信息化建设软件产品的软件公司。

Established in 1992, Hongye Technology is one of the earliest domestic companies engaged in CAD software development for engineering design. It devotes to the combination of building specialty and computer software technology and providing designers with CAD software for engineering design and software products for urban information construction.

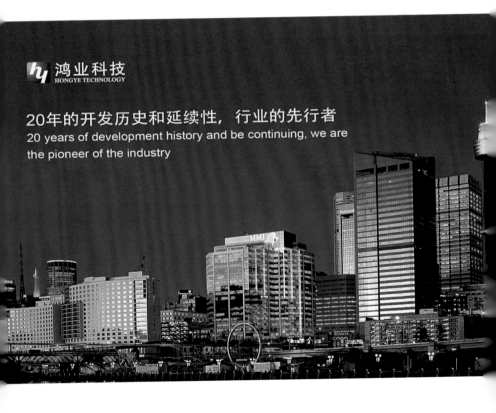

鸿业公司总部设在北京，下设上海鸿业同行信息技术有限公司和洛阳高新鸿业科技有限公司。从1992年至今，历经近二十年的风雨征程，鸿业科技已经形成专业的技术力量和庞大的销售团队。软件产品不断创新，陆续推出工程CAD系列设计软件，涵盖给水排水、暖通空调、规划总图、市政道路及管线、日照分析等建筑规划专业。从1998年起进一步推出城市规划管理信息系统，涉及规划管理信息解决方案、地理信息、控规管理、道路交通设施等系统管理软件，其中多项软件产品获得国家专业权威机构的认证和基金支持。

鸿业科技拥有用户三千余家，拥有超过一百人的研发队伍。目前，鸿业科技正在努力构建包括建筑、MEP、市政管线、市政道路在内的完整产品线。在AEC行业内，鸿业科技将为客户提供基于BIM理念的咨询服务、系统集成服务、软件定制及培训服务，帮助客户实现从二维绘图到BIM的转变。

The Company is headquartered in Beijing with branches of Shanghai Hongye Tongxing Information Technology. Co., Ltd. and Luoyang Hi-tech Hongye Technology Co., Ltd. In the two decades from 1992, Hongye Technology has established its professional technical power and a large sales team. Its software products have been constantly innovated and the CAD software for engineering design are successively developed, covering the construction planning specialty such as water supplying and drainage, HVAC (heating ventilation & air condition), general planning layout, municipal road and pipeline, sunshine analysis and etc. Since 1998, it has further developed the information system for urban planning management, involving the system management software for planning management information solutions, geographical information, regulatory planning management, road and traffic facilities and so on, many of which have been certified and funded by the national professional authorities.

Presently Hongye has more than 3000 enterprise users, More than one hundred researchers and developers in the R&D Center of Hongye are engaged in the CAD software development. Now Hongye is endeavoring to build a complete production line including building, MEP, municipal pipeline and municipal road. In AEC industry, Hongye Technology will provide customers with consulting service, system integration service, software customizing and training service based upon BIM concepts to assist them to realize the conversion from 2D drawing to BIM.

电话：（+86）010-88312202
邮箱：BIM@hongye.com.cn
网址：http://www.hongye.com.cn

Tel：（+86）010-88312202
Email：BIM@hongye.com.cn
Wel：http://www.hongye.com.cn